Electrostatics at the Molecular Level

Electrostatics at the Molecular Level

Ulrich Zürcher
Cleveland State University

Morgan & Claypool Publishers

Rights & Permissions
To obtain permission to re-use copyrighted material from Morgan & Claypool Publishers, please contact info@morganclaypool.com.

ISBN   978-1-64327-186-6 (ebook)
ISBN   978-1-64327-183-5 (print)
ISBN   978-1-64327-184-2 (mobi)

DOI   10.1088/978-1-64327-186-6

Version: 20181001

IOP Concise Physics
ISSN 2053-2571 (online)
ISSN 2054-7307 (print)

A Morgan & Claypool publication as part of IOP Concise Physics
Published by Morgan & Claypool Publishers, 1210 Fifth Avenue, Suite 250, San Rafael, CA, 94901, USA

IOP Publishing, Temple Circus, Temple Way, Bristol BS1 6HG, UK

Contents

Preface

In the words of the late Leonard Bernstein in his 1973 Norton Lecture Series at Harvard University: 'the best way to know a "thing" is in the context of another discipline.' The classical mechanics portion of introductory physics follows this paradigm, and one examines, for example, blocks on incline planes that essentially are very simplified version of problems from mechanical engineering. In particular, forces are introduced that describe the interaction of bodies on that scale: if a block sits on a (horizontal) surface, its motion is constrained $y(t) = 0$ (for some coordinate system) or, alternatively, the block is subject to a constrained or normal force F_N that opposes the weight of the block.

Elementary electrostatics lacks much of the richness of classical mechanics: electrostatics is essentially reduced to the calculation of the force on a point charge due to several other point charges at some different location. The value of the point charges and the distances are seemingly chosen 'out of thin air', and the problems are generally not put in context with another subject matter. For example, students calculate the force between two 1 µC separated by a distance of 1 mm and find that the repulsive Coulomb force is $F \simeq 10^4$ N. These values do not correspond to any realistic system so that the Coulomb force cannot be compared to any other force acting on the system; because the problem lacks a proper physical context, students cannot gain insight into the relevance of electrostatic forces.

This difficulty arises because electrostatics at the *macroscopic* level is essentially limited to 'curiosities' such as the build-up of charges when brushing hair, the 'clinging' of laundry from a dryer, and the 'electric shock' when touching a door knob after scuffing one's feet on a carpet; on the other hand, demonstrations of Coulomb's law require 'fine-tuning' of parameters. In fact, electrostatic forces went largely unnoticed until the mid-eighteenth century because macroscopic objects have zero net charge (i.e. large positive and negative charges perfectly cancel each other out).

This cancellation of positive and negative charges does not apply at the microscopic level, i.e. for atoms and molecules, so that the electrostatic forces are dominant at that level: it determines the binding of atoms in molecules and the forces between molecules as they form larger structures. Coulomb's law contributes to 'pseudo-forces' that are used to explain atomic and molecular motion. As a result of the Coulomb attraction between positive and negative charges, atoms and molecules would collapse; this is prevented by quantum-mechanical repulsion (Pauli exclusion principle). For a wide range of problems, this repulsion can be described by a repulsive force, analogous to the description of constraints for macroscopic objects in terms of 'normal forces'.

This text aims to familiarize the reader with electrostatic forces at the atomic and molecular level, that is, the 'other discipline' (in Bernstein's quote) for electrostatics is 'chemistry'. The purpose of this text is not to find the most accurate calculation of bond length and binding energies. We focus rather on a discussion of appropriate scales (for mass, time, and length, among others). The different units used for various physical quantities is often quite bewildering for non-specialists: energies are given in *kilo-joules per mol*, dipole moments in *debyes*, and polarizations in *atomic units*. Much of this work was done in the 1930s to 1960s and some of this literature is still relevant today. Reading the older literature is further complicated by the fact that Coulomb's law is expressed in cgs (centimeter-gram-second) units and the unit of electric charge is *stat-Coulomb*. Chemists only add to this confusion as they measure frequencies in units of *inverse-centimeter* (or *kayser*). While the familiar SI system provides a consistent set of units for physical quantities, its use for atomic and molecular systems is awkward since most quantities are very small (e.g. masses are of the order 10^{-26} kg and charges are of the order 10^{-19} C). We use instead the *atomic mass unit* 1 u (or *dalton* Da) as the unit of mass and the elementary charge 1 e as the unit of electric charge. In this way, all physical quantities have values of the order of unity, which greatly helps to understand the physical relevance of the various terms in the electrostatic interaction between atoms in molecules and between molecules.

We assume that the reader is familiar with basic concepts in physics and chemistry covered in an introductory physics course (either algebra or calculus based) and courses in organic and general chemistry. More advanced topics covered in upper-level physics and chemistry courses (mechanics, electricity and magnetism, thermal and statistical physics, quantum mechanics, physical chemistry, organic chemistry, biochemistry) are explained in some detail. We aim in this book to give the reader an appreciation of the order of magnitude of electric phenomena at the molecular level. We make efforts to provide the realistic number of physical quantities (e.g. atomic radii, partial charges, binding energies, etc.); however, the values of these numbers often depend on the experimental method or are somewhat model-dependent in theoretical calculations. We do not include a discussion of the uncertainty of these values and caution the reader to consult the literature to find the latest numerical values including a detailed discussion of their uncertainties.

Acknowledgments

I would like to thank students from my College Physics course at Cleveland State University, whose inquiring minds stimulated me to find applications of electro-statics to problems from chemistry that helped me to enrich the course. I learned much basic chemistry over these years from discussions with my colleague Prof John Masnovi of the Department of Chemistry, Cleveland State University, although any error reflects my own misunderstanding. I would like to thank Marisa Hollinshead for a careful reading of the manuscript and helping to make it more readable.

Author biography

Ulrich Zürcher

Ulrich Zürcher earned a PhD in Theoretical Physics from the University of Basel in Switzerland in 1989. He moved to the US and took several postdoctoral positions in Theoretical Physical Chemistry (with the late Robert Silbey at the Massachusetts Institute of Technology and Tom Keyes at Boston University) and Statistical Physics (with Charles Doering at Clarkson University, now at the University of Michigan). He then 'gravitated' towards teaching at the undergraduate level, first at the University of Rhode Island and since 2003 at Cleveland State University. He was promoted to Associate Professor in 2010 and to Professor in 2017.

Symbols

N_A	Avogadro's number (6.022×10^{23} mol^{-1})
e	elementary charge (1.602×10^{-19} C)
c	speed of light (3.00×10^8 m s^{-1})
k_B	Boltzmann constant (1.381×10^{-23} J K^{-1})
R	ideal gas constant (8.31 J mol^{-1} K^{-1})
h	Planck's constant (6.626×10^{-34} J s)
$\hbar = h/2\pi$	reduced Planck's constant (1.054×10^{-34} J s)
λ	wavelength
f	frequency
T	period
m, M	mass
(x, y, z)	three-dimensional coordinates
r, R	radius
E	energy
V	potential energy
T	kinetic energy
q	charge
δq	partial charge
p	dipole moment
α	polarizability
ϵ	permittivity
E	electric field
B	magnetic field
$\mathcal{L}$	length scale
$\mathcal{E}$	energy scale
$\mathcal{T}$	time scale

Electrostatics at the Molecular Level

Ulrich Zürcher

Chapter 1

Introduction

1.1 Coulomb forces at the macroscopic level

There are only four fundamental forces in physics. Newton's universal law of gravitation describes the attractive force between two masses m_1 and m_2 separated by a distance r_{12}: $F = Gm_1m_2/r_{12}^2$. The Coulomb force between two electric charges q_1 and q_2 at rest separated by a distance r_{12} has the same inverse-square law behavior $F_{12} = (4\pi\epsilon_0)^{-1}q_1q_2/r_{12}^2$[1]. The two remaining forces are nuclear forces: the strong force that binds nucleons (protons and neutrons in nuclei) and the weak force that describes radioactive decay. The strength of forces is characterized by a dimensionless coupling constant [14]; one finds 14.4 for the strong force, 1/137 for the electromagnetic force, 1.0×10^{-3} for the weak force, and 5.9×10^{-39} for the gravitational force. However, despite their strength the strong and weak forces rarely play a role in many physical situations since their range is limited to the range of atomic nuclei.

Both the Coulomb and gravitational force decay with the inverse-square power of the distance between masses or charges $F \sim 1/r^2$ and thus have infinite range. It is perhaps surprising that the Coulomb force is the dominant force since the Coulomb force between localized charges rarely plays a role for macroscopic objects. As any physics teacher will attest, classroom demonstrations for electrostatics are challenging and generally involve carefully chosen objects, such as lightweight styrofoam packaging 'beans' that hang on long insulating strings from the ceiling [9]. The disappearance of the Coulomb force is explained by the fact that under most circumstances the enormous positive and negative charges almost perfectly cancel out so that the net electric charge on a macroscopic object is zero and electromagnetic forces vanish.

Many forces encountered in introductory physics are based on electrostatic forces although in a form that does not resemble the Coulomb force: examples are the

---

[1] In some (introductory) texts, the abbreviation $k = (4\pi\epsilon_0)^{-1}$ is used.

normal force, the force of friction (static and kinetic), and the elastic force of a spring. These forces do not follow the $1/r^2$-dependence from Coulomb's law and their magnitudes are described by 'coefficients' that capture some properties of the object. If we take the vertical direction to be the y-direction, the normal force is determined by the condition $\sum F_y = 0$; if there are no other vertical forces acting on the block, the normal force is equal to the weight of the block, $F_N = mg$. The sliding friction $f = \mu F_N$ depends on the properties of the block and the surface via a dimensionless 'coefficient of friction' μ that, in general, has to be obtained experimentally.

This 'disguise' of the electrostatic force is familiar from the gravitational force. The gravitational force between various objects is extremely weak and can be ignored with the exception of the interaction between an object (a box, for example) with mass $m_1 = m$ and the Earth $m_2 = M_E$ separated by the Earth radius $r_{12} = R_E$ so that $F = (GM_E/R_E^2)m = mg$, where $g = 9.8$ m s^{-2} is the acceleration due to gravity; that is, the gravitational interaction is reduced to a much simpler form and is referred to as 'weight'. In fact, weight and the law of universal gravity have properties that are so different from each other that it was not obvious at all that 'earthly' objects obey the same physical laws as 'heavenly' bodies (e.g. planets) observed in the night sky.

Macroscopic realizations of electrostatic forces have properties that are counter-intuitive. It is far from obvious that sliding friction is independent of the contact area: the friction force on a metal sphere and a metal cube are the same (assuming that they are sitting on the same surface). One might think that the block with the much larger contact area experiences a much larger frictional force, which is, however, not the case. This strange property of the frictional force is explained by the roughness of a (polished) surface at the microscopic scale, i.e. atomic and molecular scales (see, for example, [13]). Positive and negative charges are separated in atoms and molecules so that 'mountain tips' on the surface and the object exert enormous (repulsive) forces on each other, so that a relatively small number of tips are responsible for both the normal and friction forces between the object and the surface.

1.2 Coulomb forces at the microscopic level

Electrostatic forces play a central role at the atomic and molecular level, although they are in a form that bears little resemblance to Coulomb force law. This is because the attractive force between electrons and protons is balanced by a repulsive force due to the quantum-mechanical nature of electrons. Several atoms assemble to form molecules and their interactions are described in terms of effective or *pseudo-forces* that describe (chemical) bonds. The energy to break up a bond is often referred to as 'chemical energy', and the connection to electrostatic energy may suggest that it is based on a different type of interaction with no relation to the forces discussed in a physics course. It turns out that positive and negative charges are separated in atoms and molecules so that attractive electrostatic forces are dominant at atomistic length scales (of the order of the size of one atom or 10^{-10} m, or one angstrom). Attractive Coulomb forces would cause the atoms and molecules to collapse; thus their stability requires that a repulsive force is present as well. Because neither the strong or weak force plays a significant role at atomistic length scales, the repulsion must be due to an intrinsic property of the involved particles. We show that the quantum-mechanical, or wave-like, properties of electrons mimic the properties of a repulsive force between atoms in molecules.

The chemical properties of elements can often be attributed to the behavior of a few valence electrons in an outer shell or missing electrons to fill an outer shell. For example, alkali atoms in Group I of the periodic table (Li, Na, K, ...) have a closed shell and one single electron in an outer shell, whereas halogen atoms in Group VII (F, Cl, Br, ...) have a closed shell except for one electron. The charge is more than plus-or-minus one elementary charge for elements in other groups of the periodic table. If the molecule is relatively small, one therefore might expect that the calculation of binding energies for molecules is straightforward and essentially reduces to finite sums of Coulomb interactions between pairs of ions. The calculation may only become more challenging in the continuum limit when the number of charges is infinite and a finite sum has to be replaced by an infinite sum or an integral. The latter case applies for macroscopic objects for which the number of charges is of the order of Avogadro's number ($N_A = 6.02 \times 10^{23}$ mol^{-1}).

However, the situation involving more than one electron is more challenging than a single-electron problem since electrons must be described as a particle wave governed by the laws of quantum mechanics rather than the laws of classical (Newtonian) mechanics. In particular, the Heisenberg uncertainty relation implies that the electrons cannot be at rest, i.e. they cannot be confined to a small region in space. On the other hand, the mass of atomic nuclei M are about three or four orders of magnitude greater than the mass of electrons m_e so that $M/m_e \sim 10^4$, and nuclei obey the laws of classical mechanics. The dynamics of electrons occur on time scales that are very short compared to the time scales of the motion of nuclei. This separation of time scales allows the electronic configurations to be calculated by assuming that the position of the atomic nuclei are fixed. This is the basis of the so-called Born–Oppenheim approximation [1].

The dynamics of molecular systems is described in a mixed fashion: electrons are described by a Schrödinger equation assuming that the positions of the ions $\{\vec{R_i}\}_i$ are fixed. The corresponding Hamiltonian for the many-electron problem depends on the coordinate of the nuclei, $\mathcal{H}(\{\vec{R_i}\})$. One finds an expectation value of the electronic energy that depends on the position of the ions $E(\{\vec{R_i}\})$. While this procedure already poses considerable challenges, finding the solution of a many-electron Schrödinger equation is much more difficult than solving a one-electron problem since the Coulomb repulsion between pairs of electrons and the 'effective' repulsion due to the Pauli exclusion principle must be included. It is therefore not surprising that the solution of the quantum-mechanical two-electron problem is considerably more challenging than finding the solution of the Schödinger problem for a one-electron system [2]. In general, finding the solution of a many-electron Schrödinger equation must be done numerically, which was made possible by the availability of computers in the 1970s. The group around Pople at Carnegie-Mellon University developed a software package, 'Gaussian'[2], that is still used today [10]. This subfield of chemistry is called 'quantum chemistry'.

This approach is not possible in a limit when the number of electrons becomes very large, as is the case for an electron gas in an external potential. In 1964, Hohenberg and Kohn developed a theory in terms of the electron density [5] and found the ground state that includes exchange and correlation effects [7]. The computational technique based on the Hohenberg–Kohn theorem, referred to as *density functional theory*, is used to predict material properties. A starting point for a review of density functional theory can be found in [6].

The positions of the ions are assumed to be fixed when the many-electron problem is solved, which implies that the position of the ions enter as parameters in the electronic Hamiltonian. This is a special case of a more general situation considered independently by Hellmann and Feynman in the late 1930s [4]. The Hamiltonian for the electron depends on a parameter λ: $H = H(\lambda)$; so that the derivative of the Hamiltonian with respect to λ: is given by

$$\frac{\partial H}{\partial \lambda} = \lim_{\Delta\lambda \to 0} \frac{1}{\Delta\lambda}[H(\lambda + \Delta\lambda) - H(\lambda)]. \tag{1.1}$$

The force for the quantum-mechanical state is then defined as the expectation value:

$$f_\lambda = -\left\langle \Psi \left| \frac{\partial H}{\partial \lambda} \right| \Psi \right\rangle = \int d^3\vec{r}\ \Psi(\vec{r})\frac{\partial H}{\partial \lambda}\Psi^*(\vec{r}). \tag{1.2}$$

When the parameter is the position of one of the atoms, $\lambda = \vec{R}_i$, the force on the ions can thus be calculated by taking the derivative of the energy with respect to the coordinates of the ions $\vec{f}_i$. Starting from the coordinates $\{\vec{R}_i\}$ at time t, one finds 'updated' positions of the ions at time $t + \Delta t$: $\vec{R}_i \to \vec{R}'_i = \vec{R}_i + (M_i\gamma)^{-1}\vec{f}_i\ \Delta t$, where γ is a relaxation constant (with units $[\gamma] = \mathrm{s}^{-1}$) and then solves the many-electron problem for the Hamiltonian $\mathcal{H}(\{\vec{R}'_i\})$, and so on. In many cases, these pseudo-forces can be approximated by much simpler semi-empirical forces, such as a linear force $f_i = -k(\vec{R}_i - \vec{R}_{i,0})$. That is, the electrostatic interaction mediated by electron clouds of ions can be modeled in terms of an elastic spring force between atoms. These forces form the basis of many-body computational methods that are used to model macromolecules such as proteins. The CHARMM (Chemistry at HARvard Macromolecular Mechanics[3]) program developed by Karplus and colleagues at Harvard University has been used for a wide range of studies examining the properties of (macro-)molecules. We direct the reader to the review by Brooks *et al* for a more detailed discussion of macromolecular modeling [3].

---

[2] See www.gaussian.com.
[3] See www.charmm.org.

1.3 Semi-empirical forces in molecular systems

Nature builds matter in an hierarchical order. On an atomistic scale, the building blocks are protons, neutrons, and electrons. Protons and neutrons form atomic nuclei, while electrons 'orbit' the nuclei. Atoms are the building blocks for molecules. For example, two hydrogen atoms form a hydrogen molecule H_2, or two hydrogen and one oxygen atom form a water molecule, H_2O. On much longer length scales, perhaps a few hundred micrometers, of the order of 10^{18} molecules aggregate and form crystals in which molecules form a regular pattern. The formation of crystals with long-range correlations is described by a thermodynamic phase transition [11]. Several crystals can arrange in even larger structures. In the case of H_2O, it has been estimated that about 100 ice crystals form a snow flake with hexagonal symmetry [8].

The structure of each length scale is associated with a different energy scale. Neutral atoms exist in the photosphere of the Sun at a temperature of about $T_{sun} \simeq 5000\,°C$, as is evident in the absorption spectrum of the Sun. However, molecules do not exist in the photosphere. In fact, thermolysis $2H_2O + heat \longrightarrow O_2 + 2H_2$ occurs at a temperature $T_{thermolysis} \simeq 2500\,°C$. Ice melts at much lower temperature $T_{melt} \simeq 0\,°C$; snow flakes melt at about the same temperature. We use the relation $E = RT\,(R = 8.3\,\mathrm{J\,(mol\,K)^{-1}})$ to find approximate binding energies, and find $E_{atom} \sim 1000\,\mathrm{kJ\,mol^{-1}}$ for the binding energy of electrons in atoms, $E_{molecule} \sim 16\,\mathrm{kJ\,mol^{-1}}$, and $E_{crystal} \sim 5\,\mathrm{kJ\,mol^{-1}}$ for the binding of water molecules in crystals.

A bond describes the interaction between two particles. For an atom, the bond acts between a nucleus and an electron, for a molecule, the (chemical) bond acts between two atoms, and for a crystal, the bond acts between molecules. Despite the differences in length and energy scales, the binding energies are explained by the balance between two forces acting on the particles: (1) the two particles have opposite electric charges so that the attractive Coulomb force acts between them, and (2) repulsive forces that reflect the quantum-mechanical nature of electrons (Heisenberg uncertainty and the Pauli exclusion principles). The balance of these two forces yields pseudo-forces that are used to characterize the force between the particles held together by the bond.

Thus, pseudo-forces explain the arrangement of atoms in molecules; on the other hand, the arrangement of atoms in molecules determines the distribution of charges in molecules and therefore the nature of pseudo-forces. Similarly, pseudo-forces explain the arrangement of molecules in crystals, and the arrangement of molecules in crystals determines the nature of pseudo-forces between molecules. That is, pseudo-forces are generated by the structure itself: we say that structures and corresponding forces are an *emergent behavior* [12].

So far, we have only discussed how bonds can be broken by adding heat, i.e. in the form of disorganized motion of bonds. However, chemical bonds in molecules and crystals can also be stretched and compressed if the particles of the bond vibrate with respect to each other. For small displacements of the particles from their equilibrium positions, pseudo-forces can be described by a spring constant κ and the masses of the two particles define a (reduced) mass μ. The vibrational frequency associated with the bond is determined by the (angular) frequency $\omega_0 = \sqrt{\kappa/\mu}$. Thus the transfer of energy into a bond can also be done by applying an electromagnetic (EM) wave with frequency Ω so that a time-varying electric field acts on the charged particles, $E(t) = E_0 \cos(\Omega t)$. If the two frequencies are identical $\Omega \sim \omega_0$, the external forcing is in resonance and the energy transfer from the EM wave to the bond is maximized. Infrared spectroscopy is used to characterize the (chemical) bonds in molecules and the binding of molecules.

This description does not apply for atomistic systems since electrons do not obey the laws of classical mechanics but rather the laws of quantum mechanics. Electron orbits are discrete so that electronic states cannot be excited with a resonance-like process. The electronic transition requires that the electron is in a superposition of ground and excited states from which one derives an expression for the frequency of the EM wave that facilitates such a transition. It follows in particular that electronic transitions in atoms correspond to the ultraviolet and visible part of the spectrum.

References

[1] Baym G 1969 *Lectures on Quantum Mechanics* (Reading, MA: Benjamin-Cummins)

[2] Bethe H A and Salpeter E E 1977 *Quantum Mechanics of One- and Two Electron Atoms* (New York: Plenum/Rosetta)

[3] Brooks B R, Brucoleri R E, Olafson B D, States D J, Swaminathan S and Karplus M 1983 CHARMM: A program for macromolecular energy, minimization, and dynamics calculations *J. Comput. Chem.* **4** 187–217

[4] Feynman R P 1939 Forces in molecules *Phys. Rev.* **56** 340–3

[5] Hohenberg P and Kohn W 1964 Inhomogeneous electron gas *Phys. Rev.* **136** B864–71

[6] Kohn W 1999 Nobel lecture: Electronic structure of matter — wave functions and density functionals *Rev. Mod. Phys.* **71** 1253–66

[7] Kohn W and Sham L J 1965 Self-consistent equations including exchange and correlation effects *Phys. Rev.* **140** A1133–8

[8] Libbrecht K 2004 Snowflake science—A rich mix of physics, mathematics, chemistry, and mystery *American Educator* Winter 2004–5

[9] Morse R 2016 Get real! - physically reasonably values for teaching electrostatics *Phys. Teach.* **54** 200–2

[10] Pople J A 1951 A theory of the structure of water *Proc. R. Soc.* A **205** 163–78

[11] Sethna J P 2006 *Statistical Mechanics: Entropy, Order Parameters, and Complexity* (New York: Oxford University Press)

[12] Strogatz S 2003 *Sync – How Order Emerges from Chaos in the Universe, Nature, and Daily Life* (New York: Hachette Books)

[13] Swartz S 2003 *Back-of-the-Envelope Physics* (Baltimore, MD: John Hopkins University Press)

[14] Walecka J D 2008 *Introduction to Modern Physics* (Singapore: World Scientific)

Chapter 2

Physical principles

2.1 Forces, potential energy, and equilibrium

The motion of an object with mass m is described by the time-dependence of the position $\vec{r}(t)$, velocity $\vec{v}(t) = d\vec{r}/dt$ and the acceleration $\vec{a}(t) = d\vec{v}/dt = d^2\vec{r}/dt^2$. If a force F acts on the object, the motion is governed by Newton's second law: $\vec{F} = m\vec{a}$. If the force is independent of time, the system is called *autonomous*; if the force depends on time, we deal with a *driven* system. If the force is conservative, the work for a closed path is zero $W = \oint \vec{F} \cdot d\vec{r} = 0$, and the force can be written as a gradient of a potential energy $V(\vec{r})$ such that $\vec{F} = -\nabla V$. We observe that the potential energy is not uniquely defined, since a change $V(\vec{r}) \rightarrow V(\vec{r}) + V_0$ yields the same force. It is often convenient to choose the constant such that $\lim_{|\vec{r}| \to \infty} V(\vec{r}) = 0$, i.e. the potential energy of the object is zero when it is far away from the origin. The sum of kinetic energy $T = \frac{1}{2}mv^2$ and potential energy V is the total energy $E = T + V$. We take the time derivative of the energy $dE/dt = m\vec{v} \cdot d\vec{v}/dt + \nabla V \cdot \vec{v} = \vec{v} \cdot [m\vec{a} + \nabla V]$; Newton's second law then implies that $dE/dt = 0$ and the total energy is constant. This is referred to as the conservation of (mechanical) energy. The linear momentum is defined as $\vec{p} = m\vec{v}$ so that Newton's second law can be written in the form $d\vec{p}/dt = \vec{F}$ [5, 7, 9].

We consider the motion of a block with mass m attached to a spring with constant k moving along the x-axis; see figures 2.1 and 2.2. The spring is at the equilibrium length when the block is the coordinate x_{eq} and the spring force vanishes. We introduce the displacement $\delta x = x - x_{eq}$; the spring force is proportional to the displacement $F = -k\delta x$, where k is the spring constant. Since $F < 0$ when $\delta x > 0$ and $F > 0$ when $\delta x < 0$, the spring force drives the block towards the equilibrium position and is therefore called a restoring force. We replace the coordinate $\delta x \rightarrow x$, and arrive at the equation of motion,

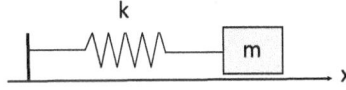

Figure 2.1. A block with mass m at the coordinate x attached to a spring with constant k. The coordinate of the block is x.

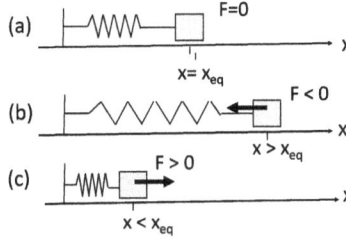

Figure 2.2. A block attached to a spring. (a) The spring is at the equilibrium position $x = x_{eq}$ and the spring force vanishes $F = 0$. (b) The spring is stretched when $x > x_{eq}$ and the spring force points towards the left, $F < 0$. (c) The spring is squeezed when $x < x_{eq}$ and the spring force points towards the right, $F > 0$.

$$a = -\frac{k}{m}x = -\omega_0^2 x, \tag{2.1}$$

where we define the angular frequency $\omega_0 = \sqrt{k/m}$. The harmonic potential is quadratic in the coordinate x,

$$V = \frac{1}{2}kx^2. \tag{2.2}$$

The motion is oscillatory with period $T_0 = 2\pi/\omega_0$ so that the motion repeats itself $x(t + n \cdot T_0) = x(t)$:

$$x(t) = A \cos(\omega_0 t + \phi); \tag{2.3}$$

$$v(t) = -v_{max} \sin(\omega_0 t + \phi) = v_{max} \cos\left(\omega_0 t + \phi + \frac{\pi}{2}\right); \tag{2.4}$$

$$a(t) = -a_{max} \cos(\omega_0 t + \phi) = a_{max} \cos(\omega_0 t + \phi + \pi); \tag{2.5}$$

where A is the amplitude, $v_{max} = A\omega_0$, and $a_{max} = A\omega_0^2$. The velocity and acceleration have the phase differences $\pi/2$ and π, respectively, with the coordinate of the block. The phase difference $\pi/2$ is a restatement of the elementary result that the speed of the block is at a maximum and minimum when the block is at the equilibrium position, $x = 0$, and at a turning point, $x = \pm A$, respectively.

If the motion is damped with a force proportional to the velocity $F_{damp} = -m\gamma v$ (where the damping constant has unit $[\gamma] = s^{-1}$), the equation of motion is given $a + \gamma v + \omega_0^2 x = 0$. If the damping is small (underdamped case), the solution is oscillatory with an amplitude that decays with time,

$$x(t) = Ae^{-t/\tau} \cos{(\omega t + \phi)}. \tag{2.6}$$

Here the time constant τ is inversely proportional to the damping constant,

$$\frac{1}{\tau} = \frac{\gamma}{2}, \tag{2.7}$$

and the period ω decreases due to damping,

$$\omega^2 = \omega_0^2 - \frac{\gamma^2}{4}, \tag{2.8}$$

and the period in the damped case is greater than in the undamped case, $T = 2\pi/\omega > T_0$.

If the damped oscillator is forced by a periodically varying force,

$$F_{\text{ext}} = F_0 \cos(\Omega t), \tag{2.9}$$

the oscillator will eventually obey the time dependence of the periodic forcing after some transient behavior,

$$x(t) = \frac{F_0}{m\omega_0^2}\chi(\Omega)\cos{[\Omega t + \phi(\Omega)]}. \tag{2.10}$$

Note that the amplitude $F_0/(m\omega_0^2) = F_0/k$ is the spring displacement when the constant force $F_{\text{ext}}(t) = F_0$ is applied. We first consider the case when the damping constant is finite $(0 < \gamma < \infty)$ and examine the dependence on the frequency of the driving force. The frequency-dependent amplitude $\chi(\Omega)$ and phase $\phi(\Omega)$ are given by

$$\chi(\Omega) = \frac{1}{\sqrt{[1 - (\Omega/\omega_0)^2]^2 + \gamma^2(\Omega/\omega_0)^2}}, \quad \phi(\Omega) = \tan^{-1}\left(\frac{\gamma\Omega/\omega_0^2}{1 - (\Omega/\omega_0)^2}\right). \tag{2.11}$$

The dependence of the amplitude and phase as a function of the scaled frequency Ω/ω_0 is shown in figure 2.3.

The limits $\Omega/\omega_0 > 1$ and $\Omega/\omega_0 < 1$ have simple explanations. (1) In the static limit $\Omega/\omega \to 0$, a constant force displaces the block to a position where the net force is zero: we find $x = F_0/k = F_0/(m\omega_0^2)$, which implies that the amplitude function is

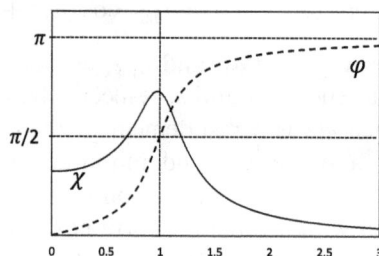

Figure 2.3. The response function χ (solid) and the phase ϕ (dashed) of a driven harmonic oscillator as a function of the scaled driving frequency Ω/ω_0, where ω_0 is the natural frequency. Resonance occurs at $\Omega = \omega_0$: the response has a maximum and the phase $\phi = \pi/2$.

equal to unity $\lim_{\Omega \to 0} \chi(\Omega) = 1$. We now consider the case when the frequency of the driving force is non-zero but small, and $\Omega < \omega_0$. The external force is approximately constant during one cycle, the block follows the external forcing, $\chi(\Omega) \simeq 1$, and the phase is close to zero $\phi(\Omega) \simeq 0$. (2) When the driving force oscillates several times during one natural period T_0 of the harmonic oscillator, the block essentially 'sees' a time-averaged external force $F(T) \to \langle F(t) \rangle = T_0^{-1} \int_t^{t+T_0} F(t')dt'$ that has a much smaller amplitude than F_0. In the limit $\Omega \gg \omega$, we have $\lim_{\Omega \to \infty} \langle F(t) \rangle = 0$ so that the amplitude function decreases, $\chi(\Omega) \sim (\omega_0/\Omega)^2 \longrightarrow 0$. Following Newton's second law, the acceleration of the object is proportional to the average force so that the block oscillates out-of phase, $\lim_{\Omega \to \infty} \phi(\Omega) = \pi$. It follows that the block moves towards the right (left) when the force $F(t)$ acts towards the left (right). When the driving frequency matches the natural frequency of the spring-block system $\Omega = \omega_0$, the block oscillates with maximum amplitude which requires that the power supplied to the block is at a maximum. Since the power is proportional to the product of force times velocity, this condition implies the velocity of the block is in phase with the driving force so that the phase follows $\phi(\Omega = \omega_0) = \pi/2$.

We now examine how the particle dynamics depends on the damping, and we are particularly interested in the case when damping is large and the motion is overdamped. We write the equations of motion as a system of coupled first-order equations (i) $dx/dt = v$ and (ii) $mdv/dt = -kx - m\gamma v$. We define the time constant $\tau = \gamma/(k/m) = \gamma m/k$ and introduce a dimensionless time $s = t/\tau$. Equation (ii) can then be written $\tau^{-1}mdv/ds = -kx - m\gamma v$. The left-hand side of the equation can be ignored for long times $t \gg \tau$ so that $0 \simeq -kx - m\gamma v$. We use equation (i) to find $0 = -kx - m\gamma dv/dt$ so that $dx/dt = (\gamma m)^{-1}F(x)$. That is, the velocity is proportional to the driving force. This result can be summarized by saying that inertia can be ignored in the high-friction limit [11].

A block attached to a spring is an example of a linear (or harmonic) oscillator. Many natural phenomena are periodic so that the motion repeats itself after a certain period $\xi(t + T) = \xi(t)$, but the underlying equation of motion does not follow a linear equation $d^2\xi/dt^2 + \omega_0^2\xi = 0$. An example is the mathematical pendulum described by the angle θ with respect to the vertical. The equation of motion is given by $d^2\theta/dt^2 + (g/L)\sin \theta = 0$ (here g is the acceleration due to gravity and L is the length of the string). The exact solution $\theta(t)$ can be written in terms of elliptic integrals [1]. In general, the nonlinear force is written in terms of a power series, e.g. $\sin \theta \simeq \theta - \theta^3/3\,!+\cdots$, and the solution can be written in terms of 'combination frequencies'; see [9]. In the limit when the amplitude of the pendulum is small $\theta < 0.5$ (recall that ϕ is measured in radians so that $\theta = 0.5\,\mathrm{rad} \simeq 30°$), we have the small-angle approximation $\sin \theta \simeq \theta$ and we arrive at the linearized form of the equation of motion $d^2\theta/dt^2 - (g/L)\theta = 0$. We recover the familiar expression for the frequency $\omega_0 = \sqrt{g/L}$.

In many situations, a linear restoring force acting on an object is the result of a competition between attractive and repulsive interactions. In many situations, the interactions are described by central forces that only depend on the radius $r > 0$. We therefore consider an object moving along the half-line $x > 0$ and subject to an attractive force $F_a(x) < 0$ and a repulsive force $F_r(x) > 0$. We assume that the repulsive force dominates near the origin, $F_r(x)/|F_a(x)| > 1$ for $x \to 0$, while the attractive force dominates for large coordinates $F_r(x)/|F_a(x)| < 1$ for $x \to \infty$. There is a point x_{eq} at which the magnitudes of the repulsive and attractive forces are equal to each other, $F_r(x)/|F_a(x)| = 1$, so that the net force on the object is zero: $F_{net} = F_r(x_{eq}) + F_a(x_{eq}) = 0$; see figure 2.4. Thus the net force has the desired property of a restoring force, namely

$$\left. \begin{array}{l} F_{net}(x) > 0 \\ F_{net}(x) < 0 \end{array} \right\} \quad \text{for} \quad \begin{cases} x < x_{eq} \\ x > x_{eq} \end{cases}. \tag{2.12}$$

We assume that the attractive and repulsive forces are conservative and define potential energies, $F_a(x) = -dV_a/dx$ and $F_r(x) = -dV_r/dx$, respectively. The potential energy of the object is given by $V(x) = V_a(x) + V_r(x)$. If the interactions vanish at large separation, it is usual to choose the potential energies such that $V(x) \to 0$ as $x \to \infty$. With this choice, an attractive (repulsive) force implies $V_a(x) < 0$ ($V_r(x) > 0$). The minimum of the potential energy defines the equilibrium coordinate at a finite distance from the origin,

$$\left. \frac{dV}{dx} \right|_{x_{eq}} = 0; \qquad x_{eq} > 0. \tag{2.13}$$

The repulsive potential dominates $V_r(x) > |V_a(x)|$ for $x < x_{eq}$, whereas the attractive potential dominates $V_r(x) < |V_a(x)|$ for $x > x_{eq}$. The second derivate is positive $d^2V/dx^2 > 0$ and defines a spring constant,

$$k = \frac{d^2V}{dx^2} > 0. \tag{2.14}$$

For small deviations from the equilibrium $\delta x = x - x_{eq}$, the harmonic approximation of the potential is then defined as $V(\delta x) = V_r(x_{eq} + \delta x) + V_a(x_{eq} + \delta x) \simeq V_h(\delta x)$, where

$$V_h(\delta x) = V_{eq} + \frac{1}{2}k(\delta x)^2. \tag{2.15}$$

The *binding energy* is defined as the energy necessary to move the object far away from the origin so that $E_b = -V_{eq}$.

It is easy to see that inverse power law behavior for the attractive and repulsive potentials, $V_a(x) \sim -1/x^i$ and $V_r(x) \sim 1/x^j$ with $i < j$, have the desired properties. They allow an analytical treatment and explain, partially at least, its popularity.

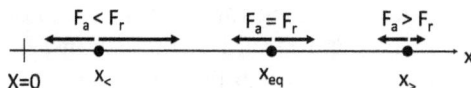

Figure 2.4. Attractive and repulsive forces $F_a(x)$ and $F_r(x)$ acting on an object moving along the half-line $x > 0$.

Best known is the Lennard-Jones potential [13]. Here, we consider an example with exponents $i = 1$ and $j = 2$:

$$V_a(x) = -V_a^0\left(\frac{a}{x}\right); \qquad V_r(x) = \frac{V_b^0}{2}\left(\frac{b}{x}\right)^2. \qquad (2.16)$$

The potentials $V_a(x)$, $V_r(x)$, and the total potential $V(x) = V_a(x) + V_r(x)$ are shown in figure 2.5. The attractive and repulsive forces follow: $F_a(x) = V_a^0 a/x^2 > 0$ and $F_r(x) = -V_b^0 b^2/x^3 < 0$. The equilibrium position x_{eq} is determined by $F_a + F_r = 0$ so that $V_b^0\, b^2/x^3 - V_a^0\, a/x^2 = (V_b^0 b^2/x - V_a^0\, a)x^{-2} = 0$, or

$$x_{eq} = \frac{V_b^0\, b^2}{V_a^0\, a}. \qquad (2.17)$$

The potential energy at the equilibrium x_{eq} follows:

$$V_{eq} = V(x_{eq}) = -\frac{(V_a^0)^2}{2V_b^0}\left(\frac{a}{b}\right)^2 < 0. \qquad (2.18)$$

The second derivative is given by $d^2V(x)/dx^2 = -2V_a^0\, b/x^3 + 3V_b^0\, b^2/x^4$ so that the spring constant is given by

$$k = \left.\frac{d^2V(x)}{dx^2}\right|_{x_{eq}} = \frac{(V_a^0)^4\, a^6}{(V_b^0)^3\, b^8} > 0. \qquad (2.19)$$

If a particle with mass m moves in the potential $V(x)$, it undergoes oscillatory motion around x_{eq} with angular frequency $\omega_0 = \sqrt{k/m}$. We find that the superposition of attractive and repulsive forces can be approximated by a spring force. While the forces F_a and F_r represent physical forces such as the electrostatic force between ions, the spring force mathematically is called a pseudo-force to distinguish it from

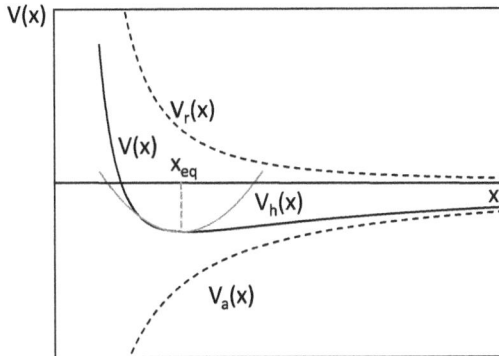

Figure 2.5. The attractive and repulsive potentials $V_a(x)$ (black, dashed) and $V_r(x)$ (black, dashed) and the total potential $V(x) = V_a(x) + V_r(x)$ (black, solid). The harmonic approximation $V_h(x)$ (red, solid) is also shown.

physical forces: there is, of course, no microscopic spring connecting the particle to its equilibrium coordinate.

2.2 Many-particle system

For a system of particles with masses $\{m_1, m_2, ...\}$ at coordinates $\{\vec{r_1}, \vec{r_2}, ...\}$, the force $\vec{F_i}$ acting on particle i is often written in terms of an external force $\vec{F}_i^{(e)}$ (e.g. due to an external field such as the Earth's gravitational field) and internal forces $\vec{F_{ij}}$ with $i \neq j$ due to the interactions between particles of the system,

$$\frac{d\vec{p_i}}{dt} = \vec{F}_i^{(e)} + \sum_{j \neq i} \vec{F_{ij}}. \tag{2.20}$$

We assume that the forces $\vec{F_{ij}}$ and $\vec{F_{ji}}$ are action–reaction pairs so that Newton's third law yields $\vec{F_{ij}} = -\vec{F_{ji}}$.

The center of mass (CoM) is defined by the average coordinate of the system (the coordinate of each particle is weighted by its mass):

$$\vec{R} = \frac{\sum_i m_i \vec{r_i}}{M}, \quad M = \sum_i m_i. \tag{2.21}$$

The motion of the CoM is independent of the internal forces and only depends on the external forces,

$$M \frac{d^2 \vec{R}}{dt^2} = \sum_i F_i^{(e)}. \tag{2.22}$$

We introduce relative coordinates $\vec{r_i} = \vec{R} + \vec{u_i}$. The linear momentum $\vec{P}$, angular momentum $\vec{L}$, and the kinetic energy T of the system follow:

$$\vec{P} = M \frac{d\vec{R}}{dt}, \tag{2.23}$$

$$\vec{L} = \vec{R} \times M \frac{d\vec{R}}{dt} + \sum_i \vec{u_i} \times m_i \frac{d\vec{u_i}}{dt}, \tag{2.24}$$

$$T = \frac{1}{2} M \left(\frac{d\vec{R}}{dt} \right)^2 + \frac{1}{2} \sum_i m_i \left(\frac{d\vec{u_i}}{dt} \right)^2. \tag{2.25}$$

As a special case, we consider a system of two bodies with masses m_1 and m_2 at the coordinates $\vec{r_1}$ and $\vec{r_2}$ interacting via a force that only depends on the difference vector $\vec{r} = \vec{r_1} - \vec{r_2}$. The CoM is then $\vec{R} = (m_1\vec{r_1} + m_2\vec{r_2})/(m_1 + m_2)$. The kinetic energy of the system can then be written as

$$T = \frac{m_1}{2}\vec{r_1}^2 + \frac{m_2}{2}\vec{r_2}^2 = \frac{M}{2}\vec{R}^2 + \frac{\mu}{2}\vec{r}^2, \qquad (2.26)$$

where the total mass is $M = m_1 + m_2$ and the *reduced* mass is defined as

$$\frac{1}{\mu} = \frac{1}{m_1} + \frac{1}{m_2}. \qquad (2.27)$$

It is often the case that one of the masses is much bigger than the other mass. For example, a hydrogen atom consists of an electron with mass m_e and a proton with mass $m_p \simeq 1836\, m_e$, i.e. the mass of the proton is about 2000 times larger than the mass of the electron. It follows that the CoM is nearly identical to the position of the proton and the reduced mass is nearly identical to the electron mass. Thus, the proton coordinate is only affected by external force (e.g. due to an external electric field), whereas the mutual interaction between proton and electron determines the orbit of the electron around the proton.

In many situations, molecules are rigid so that the distances and angles between the atoms are fixed; we say that the molecule is a *rigid body*. Thus the molecule can undergo both translational motion and rotational motion about a point $\mathcal{O}$. We use $\vec{r_i}$ for the coordinate vector of the mass m_i from $\mathcal{O}$; likewise $\vec{F_i}$ is the force on the mass m_i. The translational motion of the point $\mathcal{O}$ is described by sum of the forces $\vec{F} = \sum_i \vec{F_i}$. In some situations, the point $\mathcal{O}$ is fixed, and we are interested in the rotation about an axis $\hat{n}$ through the point $\mathcal{O}$. The torque produced by the force is given by the vector product: $\vec{F_i}$ is $\vec{\tau_i} = \vec{r_i} \times \vec{F_i}$ and the total torque is the sum $\vec{\tau} = \sum_i \vec{\tau_i}$; the torque about the axis then follows: $\tau_n = \hat{n} \cdot \vec{\tau}$. We find the distance d_i of the mass m_i from the axis $d_i = |\hat{n} \times \vec{r_i}|$ (note that $d_i = 0$ if the mass m_i lies on the axis so that $\vec{r_i}\|\hat{n}$). The moment of inertia then follows: $I = \sum_i m_i d_i^2$. We denote the angle of rotation by ϕ; the equation of motion for rotation follows:

$$I\alpha = \tau_n, \qquad (2.28)$$

where $\alpha = d^2\phi/dt^2$ is the angular acceleration. If the torque is proportional to the rotation angle, $\tau(\phi) = -\kappa\phi$, where the torsion constant has units $[\kappa] = $ Nm rad^{-1}. The equation of motion then yields $\alpha = -(\kappa/I)\phi$ so that simple harmonic motion with angular frequency $\omega = \sqrt{\kappa/I}$ follows. In some cases, the torque is derived from a potential energy $\tau = -dV/d\phi$ so that a linear restoring torque corresponds to a behavior near a local minimum ϕ_{eq}. We write $\phi = \phi_{eq} + \delta\phi$ so that the potential energy is quadratic in the angular displacement $\delta\phi$,

$$V(\delta\phi) = V_{eq} + \frac{\kappa}{2}(\delta\phi)^2, \qquad (2.29)$$

cf. equation (2.2). The equation of motion follows: $I\alpha = -dV/d\delta\phi = -\kappa\delta\phi$.

2.3 Statistical physics

Statistical physics is based on the behavior of a large number of identical particles, $N \gg 1$. We assume that each particle can be in n different 'states', each with a different energy E_ν, $\nu = 1, 2, \ldots, n$, with $E_\nu < E_{\nu+1}$ (in general, the label ν can be a continuous variable). At absolute zero temperature, $T = 0$, all particles are in the ground state with the lowest energy E_0. At non-zero temperatures, $T > 0$, statistical laws apply (Boltzmann distribution) and the average number of particles in the state ν follows:

$$p_\nu = \frac{\langle N_\nu \rangle}{N} = \frac{1}{Z} e^{-E_\nu/k_B T}, \tag{2.30}$$

where we introduce the partition function $Z = \sum_\nu e^{-E_\nu/k_B T}$ and $k_B = 1.38 \times 10^{-23}$ J K^{-1} is the Boltzmann constant [12].

For a harmonic oscillator with quadratic potential $V(x) = m\omega^2 x^2/2$, the equilibrium probability density follows a Gaussian distribution,

$$p(x) = \frac{1}{Z} \exp\left(-\frac{m\omega^2}{2k_B T} x^2\right), \tag{2.31}$$

where the partition function is given by

$$Z = \int_{-\infty}^{\infty} \exp\left(-\frac{m\omega^2}{2k_B T} x^2\right) dx = \sqrt{\frac{2\pi k_B T}{m\omega^2}}. \tag{2.32}$$

The average is zero $\langle x \rangle = 0$ and the mean-square displacement follows:

$$\langle x^2 \rangle = \int_{-\infty}^{\infty} x^2 p(x) dx = \frac{k_B T}{m\omega^2}. \tag{2.33}$$

The Gaussian probability distribution can be written $p(x) = (2\pi\langle x^2 \rangle)^{-1/2} \exp(-x^2/2\langle x^2 \rangle)$. The average harmonic oscillator potential is $\langle V \rangle = (m\omega^2/2)\langle x^2 \rangle = (m\omega^2/2) \cdot (k_B T/m\omega^2) = k_B T/2$, in agreement with the equipartition theorem.

2.4 Electromagnetic waves

Electromagnetic waves are solutions of Maxwell's equations that describe classical electrodynamics [8]. The solution of a plane wave can be written

$$\vec{E}(\vec{r}, t) = \vec{E}_0 e^{i(\vec{k}\cdot\vec{r} - \omega t)}, \tag{2.34}$$

$$\vec{B}(\vec{r}, t) = \vec{B}_0 e^{i(\vec{k}\cdot\vec{r} - \omega t)}. \tag{2.35}$$

We write $\vec{k} = k\hat{n}$, where $\hat{n}$ is the unit vector in the direction of propagation,

$$k = \sqrt{\mu\epsilon}\,\omega, \tag{2.36}$$

so that $v = 1/\sqrt{\mu\epsilon}$ is the speed of propagation. This is often written as $v = c/n$, where $c = 1/\sqrt{\mu_0\epsilon_0}$ is the speed of light in vacuum, and $n = \sqrt{\mu_r\epsilon_r}$ is the index of refraction. Since $k = 2\pi/\lambda$ and $\omega = 2\pi f$, we find $\lambda f = v$. The electric and magnetic fields are perpendicular to the direction of propagation,

$$\vec{E}_0 \cdot \hat{n} = 0 \quad \text{and} \quad \vec{B}_0 \cdot \hat{n} = 0. \tag{2.37}$$

Electromagnetic waves are *transverse*, which can be written in mathematical terms as $\vec{B}_0 = v^{-1}\hat{n} \times \vec{E}_0$. For *linear* polarized light, we can choose a coordinate system such that the wave propagates along the z-axis, $\hat{n} = \hat{e}_z$: the electric and magnetic field are directed along the x-axis and y-axis, respectively, $\vec{E}_0 = E_0\hat{e}_x$ and $\vec{B}_0 = B_0\hat{e}_y$. For circular polarized light, the electric field 'rotates' around the direction of wave propagation with frequency ω so that it has components along both the x- and y-axes:

$$E_x(z, t) = E_0 \cos(kz - \omega t), \tag{2.38}$$

$$E_y(z, t) = \mp E_0 \sin(kz - \omega t). \tag{2.39}$$

The upper sign (−) describes counterclockwise rotation (for an observer 'looking' into the incoming wave) and the electromagnetic wave is *left* circularly polarized. The lower sign (+) describes clockwise rotation (for an observer 'looking' into the incoming wave) and the electromagnetic wave is *right* circularly polarized.

The conservation of energy, momentum, and angular momentum are valid for electromagnetic waves. The flow of energy is given by the Poynting's vector $\vec{S} = \mu^{-1}\vec{E} \times \vec{B}$ and is often referred to as *intensity* with dimension energy/(area × times) = power/area. The electromagnetic momentum density associated with the electromagnetic wave is given by $\vec{g} = c^{-2}\mu^{-1}\vec{E} \times \vec{B}$ with dimension $(\text{kgm s}^{-1})\text{m}^{-3}$. The quantities $\vec{S}$ and $\vec{g}$ are proportional to each other, $\vec{g} = c^{-2}\vec{S}$. In the case of circular polarization, the wave also carries angular momentum [8].

In spectroscopy, frequencies are often expressed in units of *Kayser* or inverse centimeters $[f] = \text{cm}^{-1}$, i.e. the frequency of 1cm^{-1} corresponds to a wavelength $\lambda = 1.0$ cm. We thus arrive at the conversion: $1\text{cm}^{-1} = 3.0 \times 10^{10}$ Hz. The electromagnetic spectrum is separated into different parts. Here, the prefixes are P (Penta) for 10^{15}, T (Tera) for 10^{12}, G (Giga) for 10^9, and n (nano) for 10^{-9}. In table 2.1, we list the frequency (both in Hertz and inverse-centimeter) and wavelength of some range of the electromagnetic spectrum.

A ('black-') body at temperature T emits light over the entire spectrum, $0 < \lambda < \infty$, with a certain probability distribution, $p(\lambda, T)$, which has a maximum at a certain wavelength that depends on the temperature T (Wien's 'displacement law'), $\lambda_{\max} \simeq 2.8977 \times 10^6$ nm KT^{-1}, so that a body at room temperature emits EM waves with wavelengths $\lambda \sim 10$ μm. Visible light is emitted preferably by a body at the temperature $T \simeq 5000$ K, which is higher than the melting temperature of many materials.

Table 2.1. Properties of electromagnetic waves: frequency f in SI unit (Hz), frequency f in spectroscopic unit (kayser), and wavelength λ.

Spectrum	f	$f\,[\mathrm{cm}^{-1}]$	λ
Ultraviolet (UV)	2.5–0.79 PHz	83 300–26 300	120–400 nm
Visible	790–430 THz	26 300–14 300	400–700 nm
Infrared	430–0.3 THz	14 300–10	700 nm–1 mm
Microwave	300–0.3 GHz	10–0.01	1 mm–1 m

2.5 Quantum mechanics

2.5.1 Particles

Experiments in the beginning of the twentieth century showed that the properties of the hydrogen atom (and other atoms) cannot be explained within the laws of classical mechanics [10]. A full quantum-mechanical treatment requires solving an appropriate Schrödinger equation for the problem. In the coordinate representation, the state $|\psi\rangle$ defines the wave function $\langle \vec{r}|\psi\rangle = \psi(\vec{r})$. The time evolution of the wave function is given by the Schrödinger equation $\mathcal{H}\psi(\vec{r}, t) = [-(\hbar^2/2m)\nabla^2 + V(\vec{r})]\psi(\vec{r}, t) = i\hbar\partial\psi(\vec{r}, t)/\partial t$. The wave function $\psi(x, t)$ is the solution of the Schrödinger equation $H\psi = (i/\hbar)\partial\psi/\partial t$. We assume oscillatory time dependence is assumed with a fixed frequency $\omega = E/\hbar$ so that $\psi(\vec{r}, t) = \psi(\vec{r})$ $\exp(i(E/\hbar t))$, and arrive at the time-independent Schrödinger equation $\mathcal{H}\psi = E\psi$, where E is the energy eigenvalue. A full quantum-mechanical treatment is beyond the scope of this text and the reader is referred to standard texts [3, 6].

Heisenberg's principle states that the position and momentum of a particle cannot be determined simultaneously: $\Delta x\,\Delta p > h/4\pi$. If the particle has a fixed momentum, it corresponds to a 'matter' wave with fixed wavelength λ, $\psi(x) \sim \exp(i2\pi x/\lambda)$, and thus has infinite extension, $-\infty < x < +\infty$. The wave-like properties inherent in quantum mechanics yields another uncertainty principle relating energy and time. A wave with fixed frequency corresponds to a particle with fixed energy, $\omega = E/\hbar$. Such waves have no beginning or end: a particle wave describes a particle that can be at any point at any time. A localized disturbance corresponds to a wave packet and consists of a superposition of waves with frequencies centered around a center, $\omega_0 - \Delta\omega/2 < \omega < \omega_0 + \Delta\omega/2$. For a particle wave, this uncertainty in the frequency corresponds to an uncertainty in energy, $\Delta\omega = \Delta E/\hbar$, and we arrive at the uncertainty principle for time and energy, $\Delta t\,\Delta E > h/4\pi$.

We are primarily interested in the qualitative behavior of atoms and molecules and order of magnitude estimates of their properties, and it sufficient to use simple models and approximate solutions. We use the earliest formulation based on de Broglie's concept of matter waves. The wavelength of a particle with momentum $p = mv$ is given by ('de Broglie wavelength')

$$\lambda = \frac{h}{p}, \qquad (2.40)$$

where $h = 6.62 \times 10^{-34}$J s is Planck's constant. The frequency of the electron wave is determined by the relativistic rest energy of the electron $f = mc^2/h$ so that the phase electron follows $\lambda f = (h/mv)(mc^2/h) = c^2/v > c$ and is thus *greater* than the speed of light. The classical velocity of the electron would correspond to the group velocity of the electron [see 14].

Wave-like properties of a particle are important when the characteristic length a defined by the problem is the same order of magnitude as the de Broglie wavelength $\lambda \simeq a$. If the particle is confined to a square well with width a, it forms a standing wave with nodes at $x = 0$ and $x = a$ so that the size of the 'box' is an integer multiple of half a wavelength, $a = n\lambda/2 = n\lambda/2$, for $n = 1, 2, \ldots$ or $\lambda = 2a/n$. The momentum of the particle follows: $p = h\, n/2a$, and is independent of the mass of the particle. We set this expression equal to the (classical) momentum of a particle with mass m and velocity v $p = h\, n/2a = mv$, and conclude that the speed of the particle is inversely proportional to the mass $v \sim m^{-1}$. The state $n = 0$ is not allowed, since it would correspond to a particle at rest, $p = 0$, and would violate the Heisenberg uncertainty principle $\Delta x \Delta p \geqslant h/4\pi$. The energy of the particle follows, $E_n = p_n^2/2m$, or

$$E_n = E_0 n^2, \qquad E_0 = \frac{h^2}{8ma^2}. \qquad (2.41)$$

This model has been used to give reasonable estimates for electronic states in cyanine dyes [see 2]. An electron moves freely along the backbone of a nearly linear molecule, except near the ends when strong electrostatic forces prevent the electron from moving too far from it. The interaction is described by the 'walls' of the square-well potential. A reasonable box length is $a = 815$ pm (for a certain dye) so that the wavelengths are $\lambda_n = 2 \cdot 815$ pm$/n$ and the corresponding momenta are $p_n = 4.1 \times 10^{-25}$kgm s$^{-1} \cdot n$, and the energy is $E_n = 9.1 \times 10^{-20}J\cdot n^2$.

A quadratic potential $V(x) = (k/2)x^2$ is a good approximation for the potential to describe the dynamics of bound particles. Since the spring constant and the mass of the particle determine the frequency $\omega = \sqrt{k/m}$, we write $V(x) = m\omega^2 x^2/2$. The solution of the Schrödinger equation shows that the energy is quantized,

$$E_n = E_{0,\,\mathrm{harm}}\left(n + \frac{1}{2}\right), \qquad E_{0,\,\mathrm{harm}} = \hbar\omega. \qquad (2.42)$$

Since particles are confined both in the square well and harmonic potentials, the expressions for the energy in these two cases should yield comparable results. Thus, we set $h^2/(8ma^2) \sim h\sqrt{k/m}$, where we use the expression for the frequency ω in terms of the spring constant and the mass of the particle. We find $k \sim h^2/ma^4$ so that, in particular, the effective spring constant is inversely proportional to the mass of the particle, $k \sim m^{-1}$. If we use the size of an atom as the box length $a \sim 10^{-10}$m, we find that the 'effective' spring constant is of the order $\mathcal{O}(k) \sim 1000$N m^{-1} for electrons.

2.5.2 Electromagnetic waves

The photo effect describes the ejection of electrons from a metal surface by the exposure of light. It is found (1) that electrons are ejected as soon as the light is turned on, and (2) that there is cut-off frequency ω_c (or corresponding wavelength) such that no electrons are ejected using light with frequency $\omega < \omega_c$ no matter what intensity of light is used. These two properties cannot be explained when light is treated as a classical electromagnetic wave (i.e. the solution of Maxwell's equations in free space). The photo effect was explained by Einstein in 1905, when he postulated that the energy carried by light is transmitted in discrete packets: we say that electromagnetic *waves* have *particle*-like properties. This was aptly described by Sir William Bragg, who writes, 'It is as if one dropped a plank into the seas from a height of 100 feet, and found that the spreading ripples was able, after traveling 1000 miles becoming infinitesimal in comparison with its original amount, to act on a wooden ship in such a way that a plank of the ship flew out of its place to a height of 100 feet[1].' The 'particles' of light are referred to as photons (or γ-particles). They are particles with zero rest mass, $m_\gamma = 0$, that travel at the speed of light c. The relativistic energy–momentum relation $E^2 - (pc)^2 = m^2 c^4$ yields $E = pc$. Planck's constant yields the relation between wave- and particle-like properties:

$$E_\gamma = hf, \qquad (2.43)$$

$$p_\gamma = \frac{h}{\lambda}. \qquad (2.44)$$

Inserting into $E = pc$ yields the familiar relation between frequency and wavelength $hf = (h/\lambda) c$ or $f = c/\lambda$. We show in table 2.2 the energy and momentum of a photon corresponding to the parts of the electromagnetic spectrum in table 2.1. We conclude that the typical energy of a confined electron $\hat{E}_e$ corresponds to the energy of photons in the UV–vis part of the spectrum, whereas the energy of a confined nucleon $\hat{E}_{p, n}$ corresponds to the energy of photons in the infrared part of the spectrum. This shows that spectroscopy with visible light can be used to probe the electronic properties of atoms and molecules, whereas infrared spectroscopy must be used to explore the (relative) motion of atoms in molecules.

The spin (intrinsic angular momentum) of a photon is quantized $|\vec{S}| = \hbar\sqrt{s(s+1)}$ with $s = 1$. A spin-1 particle has three possible values of the z-component, $S_z = \pm\hbar$ and $S_z = 0$; the case $S_z = 0$ is excluded for a particle with zero mass. The state $m = +1$ ($m = -1$) describes a photon with positive (negative) *helicity* and corresponds to left (right) circularly polarized light. The intensity of an electromagnetic wave with a fixed wavelength λ can be related to the number of photons with energy $E = h c/\lambda$ crossing a unit area per unit time. Similarly, results are valid for the field-momentum density and the Maxwell stress tensor, which are related to the photon momentum and angular

---

[1] This unattributed quote is found in Holbrow C H, Lloyd J N and Amato J C 1999 *Modern Introductory Physics* (New York: Springer-Verlag) p 340.

Table 2.2. Particle-like properties of electromagnetic waves: photon energy E_γ [10^{-20}J], photon momentum p_γ [10^{-27}kgm s^{-1}] for different parts of the spectrum.

Spectrum	E_γ	p_γ
Ultraviolet (UV)	170–50	5.5–1.7
Visible	50–30	1.7–1
Infrared	30–2 $\times 10^{-2}$	1–6.6 $\times 10^{-4}$
Microwave	2×10^{-3}–2×10^{-5}	6.6×10^{-4}–6.6×10^{-7}

momentum crossing a surface. For example, an intensity 1 W m^{-2} of green light ($\lambda = 600$ nm) corresponds to 3 photons per second crossing the approximate size of an atom 1 nm × 1 nm. A photon with energy $k_B T$ has the frequency $f_T = (k_B/h)\, T$ and the wavelength $\lambda_T = (hc/k_B)T^{-1} = 1.43$ cm K$T^{-1} = 14.3 \times 10^6$ nm KT^{-1}. Thus, higher temperatures correspond to higher frequencies and lower wavelengths, which is Wien's displacement law.

A complete theory of the interaction of light and electrons involves a quantum-mechanical description of both the electron and the photon. Quantum electrodynamics is the complete theory of the emission and absorption of photons by electrons; the theory has been fully worked out and shows that the interaction has many 'strange' properties [4].

2.6 Order of magnitudes and units

Length, time, and mass are fundamental physical quantities, whose units define a system of measurements. SI units were developed to describe the behavior of macroscopic objects and reflect the human scale: length is measured in units of meters (m) and is about the length of a step, time is measured in seconds (s) and is about the time between heart beats, and mass is measured in kilograms (kg) and is about 1/100–1/50 of the mass of a typical (adult) person. These units reflect the dynamics near the Earth's surface. A block $V = 0.1$ m × 0.1 m × 0.1 m = 1 L with volume has a mass of the order of 1 kg; if it slides down a 1 m incline from rest, it acquires the speed $v = 1$m s^{-1}. The (kinetic) energy of the block is of the order of 1 J = 1 kg m^{2}s^{-2}. Acceleration $a = \Delta v/\Delta t$ is measured in units ms^{-2}, and force is measured in units of newtons 1 N = 1 kg ms^{-2}.

The connection between microscopic and macroscopic properties is made by Avogadro's number N_A, defined as the number of atoms in 12 g of carbon ^{12}C. One finds

$$N_A = 6.022 \times 10^{23}\text{mol}^{-1}. \tag{2.45}$$

This is an enormous number. For example, the number of stars in the Milky Way is of the order of 10^{11}. Since the number of galaxies in the entire (visible) universe is of the order of 10^{11}, the number of all stars in the entire Universe is of the order of

Avogadro's number. The mole is the SI unit of the amount of matter: 1 mol contains Avogadro's number of the substance. The unit of mass for molecules is the 'atomic mass unit' (amu), or Dalton, and is the characteristic scale for the mass of atoms and molecules $\mathcal{M} = 1$ u $= 1$ Da,

$$\mathcal{M} = 1.6605 \times 10^{-27} \text{kg}. \tag{2.46}$$

Because the ^{12}C atom has six neutrons and six protons, the mass of a neutron and a proton are about 1 u: $m_n = 1.008\,665$ u and $m_p = 1.007\,276$ u. The mass of the electron is about 2000 times smaller: $m_e \simeq m_p/1823$ or $m_e = 5.485\,799 \times 10^{-4}$u.

The density of (liquid) water is $\rho = 1000$ kgm^{-3} = 1.0 gcm^{-3} so that in a cube with volume $V = 1$ cm^3 there are $N = N_A/12 = 5 \times 10^{22}$ water molecules. We thus find an estimate for the volume of a single water molecule: $V' = 10^{-6}$m^3/ $5 \times 10^{22} = 2 \times 10^{-29}$m^3. We use $V' = a^3$ and find that the size of a water molecule is a fraction of a nanometer. We use 0.1 nm = 100 pm, or 1Å (angstrom), as the characteristic length scale:

$$\mathcal{L} = 1.0 \times 10^{-10} \text{m}. \tag{2.47}$$

The scaled unit of density follows: $[\rho] = \mathcal{M}/\mathcal{L}^3 = (1.66 \times 10^{-27}\text{kg})/(1.0 \times 10^{-10}\text{m})^3 =$ 1660kg m^{-3}. The volume of one mole of an ideal gas is $V_{\text{gas}} = 22.7$ L $= 22.7 \times 10^{-3}$m^3. We obtain an estimate of the 'typical' distance between atoms in gases, $d = (V_{\text{gas}}/N_A)^{1/3} = [(22.7 \times 10^{-3}\text{m}^3\text{mol}^{-1})/(6.022 \times 10^{23}\text{mol}^{-1})]^{1/3} = 3.3 \times 10^{-9}$m, or $d \simeq 33\,\mathcal{L}$.

The internal energy of a monatomic ideal gas at temperature T is given by $U = (3/2)\,n\,RT$, where $R = 8.31$ J (mol K)$^{-1}$, or, using the ideal gas law $PV = n\,RT$, $U = (3/2)\,PV$. For one mole ($n = 1$ mol), we find $U \simeq 3.8$ kJ at room temperature $T = 300$ K. The interactions between atoms and molecules in an ideal gas law are negligibly small, whereas intermolecular forces play a dominant role in the condensed phases (liquids and solids). The transformation from the solid to the liquid phase (i.e. fusion) only involves 'small' changes of the bonds between neighboring atoms/molecules associated with the rearranging of atoms/molecules. On the other hand, the transformation from liquid to gas (i.e. vaporization) involves the breaking of bonds associated with the pulling of atoms/molecules far from each other. For water, the latent heat of fusion is about one order of magnitude bigger than the latent heat of vaporization: $L_f = 33.5 \times 10^4$J kg^{-1} and $L_v = 22.6 \times 10^5$J kg^{-1}. Since 1 kg of water corresponds to 55 moles, we find $L_f = 6.1$kJ mol^{-1} and $L_v = 40.7$kJ mol^{-1}. We thus arrive at the characteristic energy scale at the molecular level,

$$\mathcal{E} = 1\,\frac{\text{kJ}}{\text{mol}} = 1.66 \times 10^{-21} \text{J}. \tag{2.48}$$

We use $(3/2)\,k_B\,T$ for the kinetic energy of a particle in thermal equilibrium at temperature T, where $k_B = R/N_A = 1.381 \times 10^{-23}$J K^{-1} is the Boltzmann constant, and find that the energy scale $\mathcal{E}$ corresponds to the temperature $T \simeq 80$ K.

Three physical quantities can be chosen to define a set of units for systems obeying Newtonian physics. In the SI system of units, one chooses mass, length, and time; alternatively, one could choose mass, length, and energy. The situation is different for microscopic systems, since they obey the laws of quantum mechanics, in which case the Heisenberg uncertainty principle provides a relation between the units of length and momentum, or between energy and time. That is, only two of the three scales ($\mathcal{M}$, $\mathcal{L}$, and $\mathcal{E}$) can be chosen independently; the remaining third scale is determined self-consistently.

We start from the scales for mass $\mathcal{M}$ and length $\mathcal{L}$ and derive the scale for energy $\mathcal{E}$. To this end, we treat the particle with mass $\mathcal{M}$ confined in a box of length $\mathcal{L}$. The expression for the energy of a particle in a box yields, cf. equation (2.41), $h^2/8\mathcal{M}\mathcal{L}^2 \simeq 3.3 \times 10^{-21}\text{J} = 3.3\ \mathcal{E}$. Alternatively, we start from the energy $\mathcal{E}$ and the mass $\mathcal{M}$, and find the momentum of the particle, $p = \sqrt{2\mathcal{M}\mathcal{E}} \simeq 2.4 \times 10^{-24}\text{kgm s}^{-1}$. The corresponding de Broglie wavelength follows, $\lambda = h/p = (6.63 \times 10^{-34}\text{J s})/(2.4 \times 10^{-24}\text{kgm s}^{-1}) \simeq 2.8 \times 10^{-10}\text{m} = 2.8\ \mathcal{L}$. Finally, we start from the scales of length $\mathcal{L}$ and energy $\mathcal{E}$. We identify the length scale with the de Broglie wavelength of a particle and find the corresponding momentum, $p = h/\mathcal{L} = (6.63 \times 10^{-34}\text{J s})/(1.0 \times 10^{-10}\text{m}) = 6.63 \times 10^{-24}\text{kgm s}^{-1}$. Since the kinetic energy of a particle with momentum p is given by $p^2/2m$, we find the mass, $m = p^2/2E = (6.63 \times 10^{-24}\text{kgm s}^{-1})^2/(2 \cdot 1.66 \times 10^{-21}\text{J}) = 13. \times 10^{-27}\text{kg} \simeq 8\mathcal{M}$.

We now turn to the characteristic timescale $\mathcal{T}$ for molecular systems. We calculate $\mathcal{T}$ from two of the three fundamental scales, $\mathcal{M}$, $\mathcal{L}$, and $\mathcal{E}$, and then show that it is compatible with the remaining scale. We first start from the scales for length $\mathcal{L}$ and the mass $\mathcal{M}$. We start from the de Broglie relation and find the momentum $p = h/\mathcal{L}$ so that the speed follows: $v = p/\mathcal{M} = h/(\mathcal{M} \cdot \mathcal{L})$. The timescale follows, $\mathcal{T} = \mathcal{L}/v = \mathcal{M}\mathcal{L}^2/h$, so that $\mathcal{T} = (10^{-10}\text{m})^2 \cdot 1.67 \times 10^{-27}\text{kg}/6.63 \times 10^{-34}\text{J s} = 2.5 \times 10^{-14}\text{s}$. We derive an energy scale from the energy–time uncertainty principle compatible with $\mathcal{E}$, $h/4\pi\mathcal{T} = (6.63 \times 10^{-34}\text{J s})/(4\pi \cdot 2.5 \times 10^{-14}\text{s}) = 2.1 \times 10^{-21}\text{J} \simeq \mathcal{E}$. Next we start from the mass $\mathcal{M}$ and $\mathcal{E}$ and find the timescale from the energy–time uncertainty principle, $\mathcal{T} = h/(4\pi \cdot \mathcal{E}) = (6.62 \times 10^{-34}\text{J s})/(4\pi \cdot 1.67 \times 10^{-21}\text{J}) \simeq 3 \times 10^{-14}\text{s}$. We find the speed of the particle from the energy and the mass, $v = \sqrt{2\mathcal{E}/\mathcal{M}} = 1000\text{m s}^{-1}$. The distance follows from speed $\times$ time, or $d = v\mathcal{T} = 1000\text{m s}^{-1} \cdot 3 \times 10^{-14}\text{s} = 3 \times 10^{-11}\text{m}$, which is compatible with the length scale $d \simeq \mathcal{L}$. Finally, we start from the length $\mathcal{L}$ and the energy $\mathcal{E}$, and find the timescale from the energy–time uncertainty principle, $\mathcal{T} = h/4\pi \cdot \mathcal{E} \simeq 3 \times 10^{-14}\text{s}$. We find the speed $\mathcal{L}/\mathcal{T} = (1 \times 10^{-10}\text{m/s})/(3 \times 10^{-14}\text{s}) = 3 \times 10^3\text{m s}^{-1}$, which is of the same order as the speed of a particle with mass $\mathcal{M}$ and energy $\mathcal{E}$, $v = \sqrt{2\mathcal{E}/\mathcal{M}} = 1000\text{m s}^{-1}$. In conclusion, we find the timescale associated with $\mathcal{M}$, $\mathcal{L}$, and $\mathcal{E}$,

$$\mathcal{T} = 1.0 \times 10^{-13}\text{s}. \tag{2.49}$$

We write $m = m'\mathcal{M}$, the length $l = l'\mathcal{L}$, energy $E = E'\mathcal{E}$, and time $t = t'\mathcal{T}$, where m', l', E', and t' are dimensionless; we refer to m' as 'mass', l' as 'length', E' as 'energy', and t' as 'time', and drop the prime. Other physical quantities are derived from these fundamental quantities. We use kinematics equations to find the scales for velocities and acceleration from $\mathcal{L}$ and $\mathcal{T}$,

$$[v] = \frac{\mathcal{L}}{\mathcal{T}} = 10^3 \, \text{ms}^{-1}, \tag{2.50}$$

so that the speed of light is written $c = 3 \times 10^5$, and for the acceleration

$$[a] = \frac{\mathcal{L}}{\mathcal{T}^2} = 1 \times 10^{16} \, \text{ms}^{-2}. \tag{2.51}$$

The unit for force follows from $\mathcal{E}$ and $\mathcal{L}$,

$$[F] = \frac{\mathcal{E}}{\mathcal{L}} = 1.67 \times 10^{-11} \text{N}, \tag{2.52}$$

or, alternatively, we use Newton's second law to derive the scale for force from the scale for $[a]$ and that for mass $\mathcal{M}, [F] = \mathcal{M}(\mathcal{L}/\mathcal{T}^2)$. Then the spring constant follows:

$$[k] = \frac{\mathcal{E}}{\mathcal{L}^2} = 0.167 \, \text{Nm}^{-1}. \tag{2.53}$$

The frequency is the inverse of the period $f = 1/T$ so that the scale for frequencies is given by

$$[f] = 1.0 \times 10^{13} \, \text{Hz} = 10 \, \text{THz}. \tag{2.54}$$

Thus, the equation relating the wavelength and frequency of an electromagnetic wave follows, $c = \lambda f$. The relationship between the energy and frequency of a photon, or alternatively the energy–time uncertainty principle, yields a relationship between the scales for energy and frequency, $[E]/[f] = \mathcal{E} \cdot \mathcal{T} = 1.66 \times 10^{-21} \text{h} \cdot 1.0 \times 10^{-13} \text{s} = 1.66 \times 10^{-34} \text{J s}$. The scales for energy and frequency are related by the uncertainty principle. We find $[E] \cdot [f] = \mathcal{E}/\mathcal{T} = (1.66 \times 10^{-21} \text{J} \cdot 1.0 \times 10^{-13} \text{s} = 1.66 \times 10^{-34} \text{J s}$. We find Planck's constant in scaled units,

$$h = 4.00, \quad \hbar = 0.634. \tag{2.55}$$

Thus the equation relating the energy and frequency of a photon follows: $E = hf$. We have $[f] = 10 \, \text{THz}$ and $\mathcal{L} = 0.1 \, \text{nm}$; the frequency and wavelength of various parts of the EM spectrum in scaled units are given in table 2.3.

In spectroscopy, frequencies are often given in units of cm^{-1}; that is, $f = 1 \, \text{cm}^{-1}$ is defined as the frequency of an electromagnetic wave with wavelength $\lambda = 1 \, \text{cm}$. We find the frequency in conventional SI units, $f = (3.0 \times 10^8 \text{m s}^{-1})/10^{-2}\text{m} = 3.0 \times 10^{10} \, \text{Hz}$, or in terms of the scaled frequency, $f = 3 \times 10^{-3} \cdot 1.0 \times 10^{13} \, \text{Hz}$. We write

$$f = 3 \times 10^{-3} \cdot \frac{f}{\text{cm}^{-1}}. \tag{2.56}$$

Table 2.3. Wave- and particle-like properties of electromagnetic waves: frequency f, wavelength λ, energy E, momentum p.

Spectrum	Frequency f	Wavelength λ	Energy E	Momentum p
Ultraviolet (UV)	250–79	1200–4000	1000–315	$3.3–1.0 \times 10^{-3}$
Visible	79–43	4000–7000	315–172	$1.0–40.6 \times 10^{-3}$
Infrared	$4 \times (10–10^{-2})$	$7 \times (10^3–10^7)$	$10^2–10^{-1}$	$6 \times (10^{-3}–10^{-7})$
Microwave	$3 \times (10^{-2}–10^{-6})$	$10^{-7}–10^{-10}$	$10^{-1}–10^{-4}$	$4 \times (10^{-7}–10^{-10})$

Here, the unit of frequency on the LHS is dimensionless (scaled) units and the unit of frequency on the RHS is cm^{-1}. For example, the vibrational frequency of sodium chloride is given as $388 \ cm^{-1}$. We find in dimensionless units,

$$f = 3 \times 10^{-3} \frac{388 cm^{-1}}{cm^{-1}} = 1.164. \tag{2.57}$$

Since Planck's constant in dimensionless units is $h = 4.00$, the energy of a photon with frequency f is now written

$$E = 1.2 \times 10^{-2} \frac{f}{cm^{-1}}. \tag{2.58}$$

The photon energy associated with the vibration of the NaCl molecule is given by

$$E = 1.2 \times 10^{-2} \frac{388 cm^{-1}}{cm^{-1}} = 4.66. \tag{2.59}$$

The relation $E = k_B T$ relates temperature to (thermal) energies. We find the conversion

$$T = 8.3 \times 10^{-3} \frac{T}{K}, \tag{2.60}$$

where temperature T on the LHS is in units of scaled energy and temperature T on the RHS is in kelvins (K). Thus room temperature $T = 300$ K corresponds to $T = 2.5$ and a typical dissociation temperature for molecules $T = 2000$ K corresponds to $T = 16.6$.

We derive the energy scale $\mathcal{E}$ from thermal properties of matter and then show that it is consistent with the (quantum-mechanical) energy of a nucleon confined to a 'box' equal to the size of an atom $\mathcal{L}$. The mass of the electron follows: $m_e/\mathcal{M} = 1/1836 = 1/\kappa$. The energy of an electron confined to $\mathcal{L}$ is inversely proportional to the mass so that for the electron, $E_e = \kappa \mathcal{E}$, or

$$E_e \simeq 1836 \ kJmol^{-1}. \tag{2.61}$$

In particular, the electron energy is much higher than the thermal energy at room temperature. We set $E_e = k_B T_e$ and find the temperature $T_e = 1834/(8.31 \times 10^{-3} K^{-1}) \simeq 220\ 000$ K. Since room temperature is much lower than T_e, we conclude that temperature effects can generally be ignored for the electronic properties of atoms and molecules.

The energy scale E_e determines the scale of quantities that describe electronic properties. The time scale of the motion of the electron T_e follows from Heisenberg's uncertainty principle, $E_e T_e = \hbar$. Since $\mathcal{E}\mathcal{T} = \hbar$, we find $\kappa \mathcal{E} T_e = \mathcal{E}\mathcal{T}$ so that

$$T_e = \frac{\mathcal{T}}{\kappa} \simeq 5.5 \times 10^{-16}\text{s}. \tag{2.62}$$

The scale for the velocity of the electron follows, $[v_e] = \mathcal{L}/T_e = \kappa[v]$; that is, electrons travel much faster than the atoms and molecules. Since momentum is given by $p = m\,v$, it follows that the momenta of electrons and atoms are the same, which is consistent with the de Broglie relation for wavelength $\lambda \simeq \mathcal{L} = \hbar/p$. The 'spring constant' for the harmonic approximation of the potential energy around a (local) minimum is proportional to the energy scale $k_e \sim E_e/\mathcal{L}^2$ so that

$$[k_e] = \kappa[k] \simeq 300 \text{ Nm}^{-1}. \tag{2.63}$$

We thus find for the (angular) frequency of oscillatory motion of an electron near a local minimum $\omega_e = \sqrt{k_e/m_e} = \sqrt{(\kappa k)/(\mathcal{M}/\kappa)} = \kappa\sqrt{k/\mathcal{M}} = \kappa\omega$, where $\omega = \sqrt{k/\mathcal{M}}$ is the frequency of the oscillatory motion of a nucleon. Since $\omega_e \sim T_e^{-1}$ and $\omega \sim \mathcal{T}^{-1}$, the scales for frequency and time are consistent with each other, as they should.

References

[1] Abramowitz M and Stegun I A 1964 *Handbook of Mathematical Functions* (New York: Dover Publications)

[2] Autschbach J 2007 Why the Particle-in-a-Box model works well for cyanine dyes but not for conjugated polyenes *J. Chem. Educ.* **84** 1840–45

[3] Eisberg R and Resnick R 1985 *Quantum Physics of Atoms, Molecules, Solids, Nuclei, and Particles* 2nd edn (New York: Wiley)

[4] Feynman R P 1994 *QED – The Strange Theory of Light and Matter* (Princeton, NJ: Princeton University Press)

[5] French A P 1971 *Newtonian Mechanics* (New York: W. W. Norton and Company)

[6] Gasiorowicz S 2003 *Quantum Physics* 3rd edn (New York: Wiley)

[7] Goldstein H 1980 *Classical Mechanics* 2nd edn (Reading, MA: Addison Wesley)

[8] Jackson J D 1999 *Classical Electrodynamics* 3rd edn (New York: Wiley)

[9] Landau L D and Lifshitz E M 1976 *Mechanics: Vol 1 of Course in Theoretical Physics* (Oxford: Butterworth and Heineman)

[10] Pais A 1986 *Inward Bound* (New York: Oxford University Press)

[11] Strogatz S 1994 *Nonlinear Dynamics and Chaos* (Boulder, CO: Perseus Press)

[12] Swendsen R H 2012 *An Introduction to Statistical Mechanics and Thermodynamics* 2nd edn (New York: Oxford University Press)

[13] Tabor D 1991 *Gases, Liquids and Solids and Other States of Matter* (New York: Cambridge University Press)

[14] Zürcher U 2016 What is the frequency of an electron wave? *Eur. J. Phys.* **37** 045401

Chapter 3

Electrostatics

3.1 Point charges

Electric charge, q, is a property of matter and is measured in the SI system in *coulombs* $[q] = C$. The electric charge is 'quantized', and its smallest unit is the elementary charge, $e = 1.609 \times 10^{-19}$ C, so that a charge can be written $q = Ne$, where N is a positive or negative integer; protons are positively charged, $q_p = e$, electrons are negatively charged, $q_e = -e$, and neutrons have no charge, $q_n = 0$. The existence of charge is evident by the forces they exert on each other. The force between two point charges q_1 and q_2 is determined by Coulomb's law: $\vec{F}_{12}$ ($\vec{F}_{21}$) is the force on the charge q_1 (q_2) due to the charge q_2 (q_1),

$$\vec{F}_{12} = -\vec{F}_{21} = \frac{1}{4\pi\epsilon_0} \frac{q_1 \cdot q_2}{r_{12}^2} \hat{r}_{12}, \tag{3.1}$$

where $\hat{r}_{12}$ is the unit vector point from charge q_1 to charge q_2; cf. figure 3.1. The forces are equal and opposite by virtue of Newton's third law. Here, ϵ_0 is the electric permeability of free space,

$$\epsilon_0 = 8.85 \times 10^{-12} \, \text{C}^2 \, \text{N}^{-1} \, \text{m}^{-2}. \tag{3.2}$$

The Coulomb constant is sometimes written in terms of the constant $k = (4\pi\epsilon_0)^{-1} \simeq 9.0 \times 10^9 \, \text{N} \, \text{m}^2 \, \text{C}^{-2}$. The Coulomb force is repulsive for like charges $q_1 > 0$ and $q_2 > 0$ or $q_1 < 0$ and $q_2 < 0$ and is attractive between unlike charges $q_1 > 0$ and $q_2 < 0$ or $q_1 < 0$ and $q_2 > 0$. In the case of multiple charges $\{q_i\}_{i=1, 2, ..., N}$ at the positions $\{\vec{r}_i\}$, the contributions from charges q_j with $j \neq i$ are given by $\vec{F}_{ij} = (4\pi\epsilon_0)^{-1} q_i q_j / r_{ij}^2 \cdot \hat{r}_{ij}$. The force on the charge q_i is given by a vector sum,

$$\vec{F}_i = \sum_{j \neq i} \vec{F}_{ij} = \frac{1}{4\pi\epsilon_0} \sum_{j \neq i} \frac{q_i q_j}{r_{ij}^2} \hat{r}_{ij}. \tag{3.3}$$

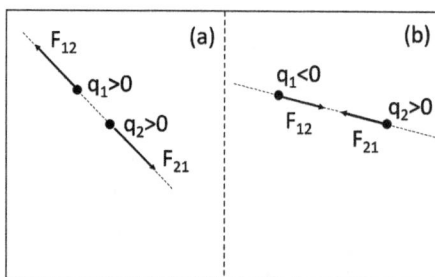

Figure 3.1. (a) The repulsive forces F_{12} and F_{21} between like charges q_1 and q_2; (b) the attractive forces $\vec{F}_{12}$ and $\vec{F}_{21}$ between unlike charges q_1 and q_2.

This is referred to as the *superposition principle*.

The force $\vec{F}_i$ on the charge q_i can be written in terms of the electric field $\vec{F}_i = q_i\vec{E}_i$ so that $\vec{E}_i = (4\pi\epsilon_0)^{-1}\sum_{j\neq i}(q_j/r_{ij}^2)\,\hat{r}_{ij}$. The RHS is defined even if there is no charge at the point $\vec{r}_i$. We define an electric field at any point $\vec{r}$ by

$$\vec{E}(\vec{r}) = \frac{1}{4\pi\epsilon_0}\sum_i \frac{q_i}{|\vec{r}-\vec{r}_i|^2}\frac{\vec{r}-\vec{r}_i}{|\vec{r}-\vec{r}_i|}, \tag{3.4}$$

where $(\vec{r}-\vec{r}_i)/|\vec{r}-\vec{r}_i|$ is a unit vector from the point $\vec{r}_i$ to the point $\vec{r}$. The unit of the electric field is $[E] = \text{N/C}$.

For a single charge at the origin, the magnitude of the electric field at the point $\vec{r}$ from the charge is given by $E = \epsilon_0^{-1}q/(4\pi r^2) = \epsilon_0^{-1}(q/A)$, where $A = 4\pi r^2$ is the surface area of a sphere with radius r. We thus have the surface integral $\int \vec{E}\cdot d\vec{A} = EA = q/\epsilon_0$. That is, the 'flux' of the electric field through the surface of the sphere is independent of the radius of the sphere and only depends on the charge inside the sphere. It can be shown that this result is correct for any static electric field, $\vec{E}$, and any volume, V, with surface ∂V. This is Gauss' law,

$$\int_{\partial V} \vec{E}\cdot d\vec{A} = \frac{1}{\epsilon_0}\sum_{\in V} q_i, \tag{3.5}$$

where $\sum_{\in V}q_i$ is the sum of the charges inside the volume V. We note that electric flux is independent of the *distribution* of the point charges inside the volume. This property is a consequence of the inverse-square behavior of the electric field $|\vec{E}| \sim r^{-2}$ [5].

Electric fields are often visualized using *electric field lines*. Electric field lines start at positive charges (or infinity) and end at negative charges (or infinity). They do not cross so that there is only one field line going through any point $\vec{r}$. The direction of the electric field $\vec{E}(\vec{r})$ is then determined by the direction of the tangent line to the electric field line. The 'density' (number of lines per unit area) is a measure of the strength of the electric field. The number of electric field lines starting and ending at positive and negative charges, respectively, is proportional to the magnitude of the

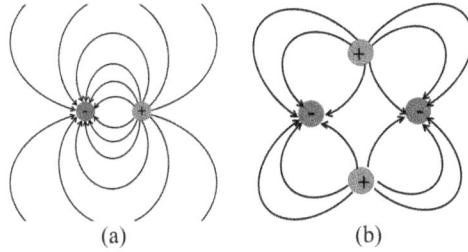

Figure 3.2. Electric field lines for (a) an electric dipole (one positive and one negative charge of equal magnitude) and (b) an electric quadrupole (two positive and two negative charges of equal magnitude).

charges. The electric field lines for a dipole and quadrupole are shown in figures 3.2(a) and (b).

We assume that some distribution of point charges produces a static electric field $\vec{E}(\vec{r})$. A charge, q, is placed at some point $\vec{r_i}$ and 'we' move the charge very slowly to the point $\vec{r_f}$. Because the electric force $\vec{F} = q\vec{E}$ acts on the charge, the electric field does work on the charge. One finds that the work is independent of the path $W_{el} = q\int_{\#1}\vec{E}\cdot d\vec{s} = q\int_{\#2}\vec{E}\cdot d\vec{s}$ and only depends on the initial and final position $W_{el} = W_{el}(\vec{r_i}, \vec{r_f})$, cf. figure 3.3. This implies that the work done by the electric field on a closed path is zero $q\oint\vec{E}\cdot d\vec{s} = 0$. We say that the electric force is *conservative* and can be derived from an (electrostatic) potential energy V [7],

$$\vec{F}(\vec{r}) = -\nabla V, \tag{3.6}$$

so that in Cartesian coordinate systems $V = V(x, y, z)$ and $F_x = -\partial V/\partial x$, $F_x = -\partial V/\partial y$, and $F_z = -\partial V/\partial z$. For two point charges q_1 at $\vec{r_1}$ and q_2 at $\vec{r_2}$, we find

$$V = \frac{1}{4\pi\epsilon_0}\frac{q_1 q_2}{r_{12}} = \frac{1}{4\pi\epsilon_0}\frac{q_1 q_2}{\sqrt{(x_1 - x_2)^2 + (y_1 - y_2)^2 + (z_1 - z_2)^2}}. \tag{3.7}$$

The potential energy is defined up to a constant value; in equation (3.7) the constant is chosen such that $V = 0$ when the two charges are far apart $r_{12} \rightarrow \infty$. For a system of charges $\{q_i\}$, the potential energy of the charge q_i follows:

$$V_i = \frac{1}{4\pi\epsilon_0}\sum_{j\neq i}\frac{q_i q_j}{r_{ij}}. \tag{3.8}$$

The total potential energy of the charge distribution follows by taking the sum $V = \sum_i V_i$, such that any pair (q_i, q_j) only contributes once to the total potential energy of the system,

$$V = \frac{1}{4\pi\epsilon_0}\sum_{i<j}\frac{q_i q_j}{|\vec{r_i} - \vec{r_j}|} = \frac{1}{8\pi\epsilon_0}\sum_{i\neq j}\frac{q_i q_j}{|\vec{r_i} - \vec{r_j}|}. \tag{3.9}$$

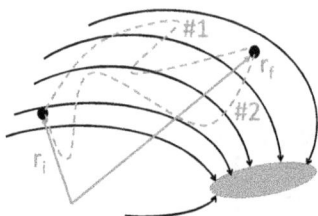

Figure 3.3. A particle with charge q is placed in a region with (static) electric field $\vec{E}$ and is moved from the initial position $\vec{r_i}$ to the final position $\vec{r_f}$. The work done by the electric field is the same along path #1 and #2: $\int_{\#1} q\vec{E} \cdot d\vec{s} = \int_{\#2} q\vec{E} \cdot d\vec{s}$.

The potential energy V_i of the charge q_i can be written in terms of an electrostatic potential $V_i = q_i\, \Phi_i$, where $\Phi_i = (4\pi\epsilon_0)^{-1}\sum_{j\neq i}(q_j/r_{ij})$. The RHS is defined even if there is no charge q_i at the point $\vec{r}$. We define the electrostatic potential at any point $\vec{r}$:

$$\Phi(\vec{r}) = \frac{1}{4\pi\epsilon_0}\sum_i \frac{q_i}{|\vec{r}-\vec{r_i}|} = \frac{1}{4\pi\epsilon_0}\sum_i \frac{q_i}{\sqrt{(x-x_i)^2 + (y-y_i)^2 + (z-z_i)^2}}. \quad (3.10)$$

The SI unit for the electric potential is *volts*, $[\Phi] = \mathrm{J\,C^{-1}} = \mathrm{V}$. The electric field is determined by the steepest descent of the potential, or, mathematically, as the gradient of Φ:

$$\vec{E} = -\nabla\Phi, \quad (3.11)$$

which is the general form of the corresponding expression in Cartesian coordinates, $E_x = -\partial\Phi/\partial x$, $E_y = -\partial\Phi/\partial y$, and $E_z = -\partial\Phi/\partial z$. Equation (3.11) is the generalization of the relation between the voltage drop ΔV and a uniform (constant) electric field E, $\Delta V = Ed$, where d is the distance between the plates of a parallel-plate capacitor; it yields another unit for the electric field, $[E] = \mathrm{N\,C^{-1}} = \mathrm{V\,m^{-1}}$.

We note that the expression equation (3.4) for the electric field requires that three sums are evaluated (one for each spatial direction) whereas the expression for potential equation (3.8) only involves a single sum. In general, the evaluation of sums with many terms is challenging, such that the calculation of the electric field is much more challenging than the calculation of the potential. Therefore, one generally calculates $\Phi(r)$ and then calculates the electric field from equation (3.11).

3.2 Continuous charge distribution

In a typical 'engineering' application, a metal sphere of radius $R = 10\,\mathrm{cm}$ is 'charged' by applying a voltage $V = 1\,\mathrm{kV}$ to the sphere. The charge on the sphere follows $Q = 4\pi\epsilon_0 Vr = 4\pi \cdot 8.85 \times 10^{-12}(\mathrm{C^2\,m^{-2}}) \cdot 10^3\,\mathrm{V} \cdot 10^{-1}\mathrm{m} \simeq 10^{-7}\mathrm{C}$; the number of elementary charges follows $N = Q/e = (10^{-7}\mathrm{C})/(1.6 \times 10^{-19}\,\mathrm{C}) \simeq 10^{12}$. This is an enormous number (a billion times a billion), and the summation over charges is replaced by an integration over a continuous charge distribution.

Macroscopic objects are differentiated into conductors (metals) and insulators (wood, rubber, etc.). For a conductor, the charges (mostly electrons) move freely and rearrange until the electron field inside the metal is zero; thus, the charges reside on the surface and are characterized by a surface charge density σ. The surface is an equipotential surface so that the component of the electric field parallel to the surface is zero, $E_{\parallel} = 0$. The perpendicular component is discontinuous, $\Delta E_{\perp} = \sigma/\epsilon_0$.

This case is of little interest in the context of this book; instead, some of the properties of atoms and molecules can be derived using an *insulator* as a model. The charges are (largely) fixed inside an insulator and define a volume charge density ρ (with unit $[\rho] = \mathrm{Cm}^{-3}$). The electrostatic potential is then written:

$$\Phi(\vec{r}) = \frac{1}{4\pi\epsilon_0} \int \frac{\rho(\vec{r}\,')}{|\vec{r}-\vec{r}\,'|}d^3\vec{r}\,'. \tag{3.12}$$

While the evaluation of the integral on the RHS can be tedious, it is conceptually straightforward. When the charge distribution has a simple geometry, e.g. rods, sphere, etc., the integral can be evaluated explicitly [4–6].

Here we discuss the example of an insulating hollow spherical shell with inner radius R_1 and outer radius R_2; this is shown in figure 3.4. The charge density follows: $\rho = (3/4\pi)Q(R_2^3 - R_1^3)^{-1}$ (Q is the total charge on the insulator). The charge density depends only on the radius r and is thus spherically symmetric so that the electric field has only a radial component, $\vec{E} = E_r\hat{r}$. Because the charge inside the sphere $r < R_1$ is zero, an application of Gauss' law shows that the electric field is zero in the hollow part:

$$E_r(r) = 0; \qquad r < R_1. \tag{3.13}$$

Inside the insulator, for $R_1 < r < R_2$, the enclosed charge is $Q_{\mathrm{in}} = Q(r^3 - R_1^3)/(R_2^3 - R_1^3)$. Since the surface area of the sphere of radius r is $A = 4\pi r^2$, the electric field follows:

$$E_r(r) = \frac{Q}{4\pi\epsilon_0} \frac{r - R_1^3/r^2}{R_2^3 - R_1^3}; \qquad R_1 < r < R_2. \tag{3.14}$$

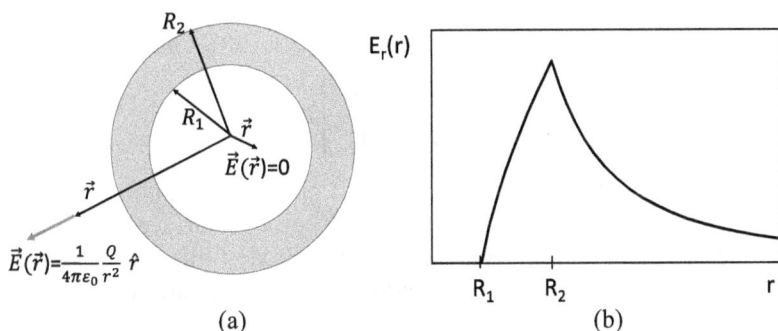

Figure 3.4. A uniformly charged hollow sphere with inner radius R_1 and outer radius R_2. The electric field is zero for $0 < r < R_1$ and $E_r(r) = (4\pi\epsilon_0)^{-1}(Q/r^2)$ for $r > R_2$.

Outside the spherical shell, the enclosed charge is equal to the total charge Q and the electric field is identical to that produced by a point charge at the center of the spherical shell:

$$\mathsf{E}_r(r) = \frac{1}{4\pi\epsilon_0}\frac{Q}{r^2}; \qquad r > R_2. \tag{3.15}$$

In the special case when the inner radius is zero $R_1 = 0$ and $R_2 = R$, we recover the case of a uniformly charged sphere. We find

$$\mathsf{E}_r(r) = \frac{Q}{4\pi\epsilon_0}\frac{r}{R^3}, \qquad r < R, \tag{3.16}$$

and $\mathsf{E}_r(r) = Q/(4\pi\epsilon_0 r^2)$ for $r > R$.

The so-called 'classical electron' is an interesting case. It assumes that the electron is a sphere with radius r_e and a charge density distribution $\rho_e(r)$ for $r < r_e$. We consider a fixed radius $0 < r < r_e$; the charge inside the radius $0 < r' < r$ generates an electric field that is 'felt' by the electric charge outside the radius $r < r' < r_e$. In a sense, the charge of the electron interacts with itself. A careful discussion shows that the self-energy $E_{\text{self}} = \kappa(4\pi\epsilon_0)^{-1}e^2/r_e$, where κ is a numerical factor of the order of unity that depends on details of the charge distribution. We use $\kappa = 1$, and assume that the rest mass of the electron is equal to the electron self-energy; we find $m_e c^2 = (4\pi\epsilon)^{-1}e^2/r_e$. The 'classical' electron radius follows: $r_e = e^2/(4\pi\epsilon_0 m_e c^2)$. We find the numerical value, $r_e = 2.81 \times 10^{-15}$m.

3.2.1 Poisson equation

For a continuous charge density, ρ, the charge enclosed in a small volume, ΔV, can be written: $Q = \rho\Delta V$. The electric flux through the surface of the small volume $\partial\Delta V$ defines the 'divergence' of the electric field, $\lim_{V\to 0}V^{-1}\int_{\partial V}\vec{E}\cdot d\vec{A} = \nabla\cdot\vec{E}$. We find

$$\nabla\cdot\vec{E} = \frac{\rho}{\epsilon_0}. \tag{3.17}$$

In Cartesian coordinates, the divergence is given by $\nabla\cdot\vec{E} = \partial E_x/\partial x + \partial E_y/\partial y + \partial E_z/\partial z$. Since the electric field can be written as the (negative) gradient of the potential, $\vec{E} = -\nabla\Phi$, we find

$$\nabla^2\Phi = \left[\frac{\partial^2}{\partial x^2} + \frac{\partial^2}{\partial y^2} + \frac{\partial^2}{\partial z^2}\right]\Phi = -\frac{\rho}{\epsilon_0}. \tag{3.18}$$

This is an inhomogeneous second-order partial differential equation; its solution is specified by the values of the potential Φ_1 and Φ_2 on two surfaces $f_{1,2}(x, y, z) = 0$. If the charge density depends only on a single coordinate (taken to be x), the partial differential equation simplifies to a second-order ordinary differential equation,

$$\frac{d^2\Phi}{dx^2} = -\frac{\rho(x)}{\epsilon_0}, \tag{3.19}$$

and the solution is specified by two boundary conditions, $\Phi(x_1) = \Phi_1$ and $\Phi(x_2) = \Phi_2$.

In equations (3.18) and (3.19) the electrostatic potential Φ is calculated from a given charge density ρ. In general, the charge density can depend on the potential so that ρ and Φ must be determined self-consistently. As an example, we discuss the distribution of positive and negative ions in an electrolytic solution near a suspended colloidal particle, shown in figure 3.5. We follow the treatment in volume II of [2], and consider the dependence along x only, $\rho = \rho(x)$ and $\Phi = \Phi(x)$. The charge density in the electrolyte is determined by the particle densities of positive and negative ions:

$$\rho(x) = en_+(x) + (-e)n_-(x). \tag{3.20}$$

We assume that the colloidal particle is fixed at the origin $x = 0$ so that $V_+ = e\Phi$ and $V_- = -e\Phi$ are the potential energies of the ions in the presence of the potential Φ. The number density of positive and negative ions then follows,

$$n_+(x) = n_0 e^{-e\Phi(x)/k_BT}, \qquad n_-(x) = n_0 e^{+e\Phi(x)/k_BT}, \tag{3.21}$$

where n_0 is an equilibrium concentration. It follows that the total charge density is given by

$$\rho(x) = en_0\left[\exp\left(-\frac{e\Phi(x)}{k_BT}\right) - \exp\left(+\frac{e\Phi(x)}{k_BT}\right)\right]; \tag{3.22}$$

thus the charge distributions of ions is *negative* near the positively charged colloidal particle. We arrive at the differential equation for the potential,

$$\frac{d^2\Phi}{dx^2} = -\frac{en_0}{\epsilon_0}\left[\exp\left(-\frac{e\Phi(x)}{k_BT}\right) - \exp\left(+\frac{e\Phi(x)}{k_BT}\right)\right]; \tag{3.23}$$

this is a nonlinear ordinary differential equation and is difficult to solve.

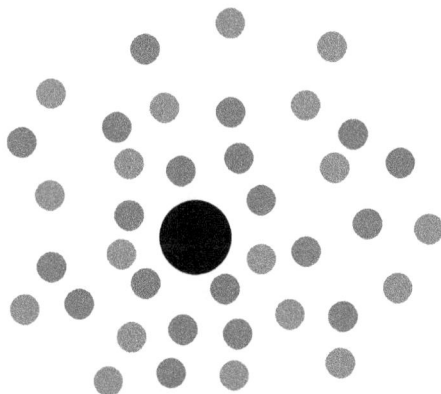

Figure 3.5. A positively charged colloidal particle (black) surrounded by positive ions (red) and negative ions (red).

We assume that the electrostatic potential energy of the ions is small compared to the thermal energy $e\Phi < k_B T$; the argument of the exponential function is less than unity, $|z| < 1$ and $\exp(\pm z) \simeq 1 + z$. We find the linearized equation:

$$\frac{d^2\Phi}{dx^2} = \frac{2e^2 n_0}{k_B T \epsilon_0} \Phi. \tag{3.24}$$

This is a Poisson equation for the potential Φ. The RHS defines the Debye length:

$$\lambda_D = \sqrt{\frac{k_B T \epsilon_0}{2 n_0 e^2}}. \tag{3.25}$$

We use the concentration $n_0 = 1$ M (i.e. one mole per liter) or $n_0 = 6.02 \times 10^{26} \mathrm{m}^{-3}$ so that at room temperature $T = 300$ K,

$$\lambda_D = \sqrt{\frac{1.38 \times 10^{-23} \mathrm{J\ K}^{-1} \cdot 300\ \mathrm{K} \cdot 8.85 \times 10^{-12} \times \mathrm{C}^2 \mathrm{N}^{-1} \mathrm{m}^{-2}}{2 \cdot 6.02 \times 10^{26} \mathrm{m}^{-3} \cdot (1.602 \times 10^{-19} \mathrm{C})^2}} = 3.4 \times 10^{-10} \mathrm{m}, \tag{3.26}$$

or $\lambda_D = 0.34\ \mathcal{L}$. We note that the Debye length is inversely proportional to the square of the concentration $\lambda \sim 1/\sqrt{n_0}$ so that for the concentration $n_0 = 10^{-2}\mathrm{M} = 10$ mM the Debye length scale is $\lambda_D = 3.4 \times 10^{-10}\mathrm{m} = 3.4\ \mathcal{L}$. We introduce the scaled Debye length,

$$\Lambda_D = \frac{\lambda_D}{\mathcal{L}}, \tag{3.27}$$

and write $x' = x/\mathcal{L}$ and arrive at

$$\frac{d^2\Phi}{dx^2} = \frac{1}{\Lambda_D^2} \Phi, \tag{3.28}$$

where we replace x' by x, $x' \to x$. The Poisson equation has two solutions, $\exp(-x/\Lambda_D)$ and $\exp(+x/\Lambda_D)$. Since the potential vanishes far from the colloidal particle, $\Phi(x) \to 0$ for $x \to \infty$, we find

$$\Phi(x) = \Phi_0 e^{-x/\Lambda_D}. \tag{3.29}$$

Since the Debye length is typically of the same order as the size of an atom, we conclude that the distribution of positive and negative ions are in equilibrium, and the total charge in a small volume ΔV is zero, except for a few layers of ions around the colloidal particle. The value of the potential at the colloidal particle is determined by the surface charge density on the colloidal particle, $\sigma = \epsilon_0 E |_{x=0} = \epsilon_0(-d\Phi/dx)_{x=0} = \epsilon_0(\Phi_0/\Lambda_D)$, so that $\Phi_0 = \sigma \Lambda_D/\epsilon_0$, and

$$\Phi(x) = \frac{\sigma \Lambda_D}{\epsilon_0} e^{-x/\Lambda_D}. \tag{3.30}$$

3.3 Multipole expansion

A 'typical' problem in electrostatics involves a molecule with 'size' a that is charged by point charges $\{q_i\}$ at positions $\{\vec{r}'_i\}$ with $r_i' < a$, and we are interested in the electric field at a point $\vec{r}$ far from the molecule $r > r_i$; see figure 3.6. The generalization to a continuous charge density $\rho(\vec{r}')$ is straightforward. It follows that the potential obeys the Laplace equation, $\nabla^2\Phi = 0$. We seek a solution in the far field and therefore have the boundary condition $\Phi(r) \to 0$ for $r \to \infty$.

It is convenient to use *spherical* polar coordinates (r, θ, ϕ) instead of Cartesian coordinates (x, y, z); we refer the reader to [1] or any text on mathematical methods in science and engineering. The Laplace equation is 'separable' in spherical coordinates (r, θ, ϕ) [5]; that is, the potential can be written as a product,

$$\Phi(r, \theta, \phi) = \frac{f_1(r)}{r}f_2(\theta)f_3(\phi). \tag{3.31}$$

The partial differential equation is transformed into a system of three coupled ordinary differential equations:

$$\frac{d^2f_3}{d\phi^2} + m^2f_3 = 0, \tag{3.32}$$

$$\frac{1}{\sin\theta}\frac{d}{d\theta}\left(\sin\theta\frac{df_2}{d\theta}\right) + \left[l(l+1) - \frac{m^2}{\sin^2\theta}\right]f_2 = 0, \tag{3.33}$$

$$\frac{d^2f_1}{dr^2} - \frac{l(l+1)}{r^2}f_1 = 0, \tag{3.34}$$

where (l, m) are *separation constants* that need to be determined from boundary conditions (here we use the standard notion and note that m must not be confused with mass).

We recall that any second-order differential equation has two linearly independent solutions. For the radial equation, the solutions are power laws, $f_1(r) \sim r^{l+1}$ and $f_1(r) \sim r^{-l}$. Since we seek a solution that remains finite for $r \to \infty$, the solution r^{l+1} is unphysical and find

$$f_1(r) = c_l r^{-l}, \tag{3.35}$$

so that the radial part decays faster for higher values of l.

For $m = 0$, the solution of equation (3.33) is $f_3(\phi) = $ const so that the solution is independent of the azimuthal angle ϕ, and this case yields the electrostatic potential

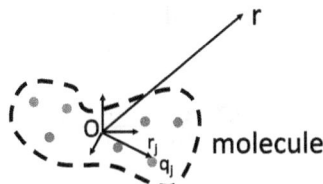

Figure 3.6. A positively charged colloidal particle (black) surrounded by positive ions (red) and negative ions (red).

for a charge distribution that is independent with respect to rotation about the 'polar axis'. For $m \neq 0$, the solutions are the familiar trigonometric functions $f_3(\phi) = \sin(m\phi)$ and $f_3(\phi) = \cos(m\phi)$. We note that the trigonometric functions have $2m$ zeros for $0 \leqslant \phi < 360°$, Alternatively, we use the Euler equation $[\cos z + i \sin z = e^{iz}]$ and find the solution $f_3(\phi) = e^{\pm im\phi}$.

For the latitudinal dependence, i.e. the dependence with respect to θ, we first consider the case $m = 0$ and write $x = \cos\theta$. One finds

$$\frac{d}{dx}\left[(1 - x^2)\frac{df_2}{dx}\right] + l(l + 1)f_2 = 0. \tag{3.36}$$

It is shown that this differential equation has a solution only if l is zero or a positive integer, $l = 0, 1, 2, \ldots$. The regular solution is the Legendre polynomial, $f_2(\theta) = P_l(\cos\theta)$, and is a polynomial of order l,

$$P_0(x) = 1, \tag{3.37}$$

$$P_1(x) = x, \tag{3.38}$$

$$P_2(x) = \frac{1}{2}(3x^2 - 1), \tag{3.39}$$

$$P_3(x) = \frac{1}{2}(5x^3 - 3x); \tag{3.40}$$

in general $P_l(x) = (2^l l!)^{-1}(d^l/dx^l)(x^2 - 1)^l$. The first six Legendre polynomials are shown in figure 3.7. Since the Legendre polynomial has l- zeros $\{\xi_i\}_{i=1,\,2,\,\ldots,\,l}$ with $P_l(\xi_i) = 0$, the θ-dependence becomes more 'oscillatory' for higher values of l. The irregular solution of equation (3.36) is unphysical. For $m \neq 0$, the equation (3.33) has a solution only for $m = -l, -l + 1, \ldots, -1, 0, 1, \ldots, l - 1, l$, and is given by the associated Legendre polynomial $P_l^m = (-1)^m(1 - x^2)^{m/2}(d^m/dx^m)P_l(x)$. We note that $P_l^m(\cos\theta)$ has $l - m$ zeros. The associated Legendre polynomials for $l = 4$ are shown in figure 3.8.

We conclude that the angular dependence of the potential at a fixed distance r_0, $\Phi(r = r_0, \theta, \phi)$, has a total number of $l + |m|$ zeroes. That is, the potential has more angular 'features' for larger values of l and $|m|$; the maximum number of zeroes $\Phi(r = r_0, \{\theta_\nu, \phi_\nu\}_\nu) = 0$ is $2l$. It is standard to introduce spherical harmonic functions, $Y_{lm}(\theta, \phi)$, that are products of associated Legendre functions and trigonometric functions [5]. As such, the potential in the 'far field', $r > a$, can be written as a superposition; this is referred to as the *multipole expansion*:

$$\Phi(r, \theta, \phi) = \frac{1}{4\pi\epsilon_0} \sum_{l=0}^{\infty} \sum_{m=-l}^{l} \frac{4\pi}{2l + 1} q_{lm} \frac{Y_{lm}(\theta, \phi)}{r^{l+1}}. \tag{3.41}$$

The coefficients q_{lm} are the 'multipole' moments and are determined by distribution of charges $\{q_i\}$ on the molecule. The moment q_{lm} has dimensions

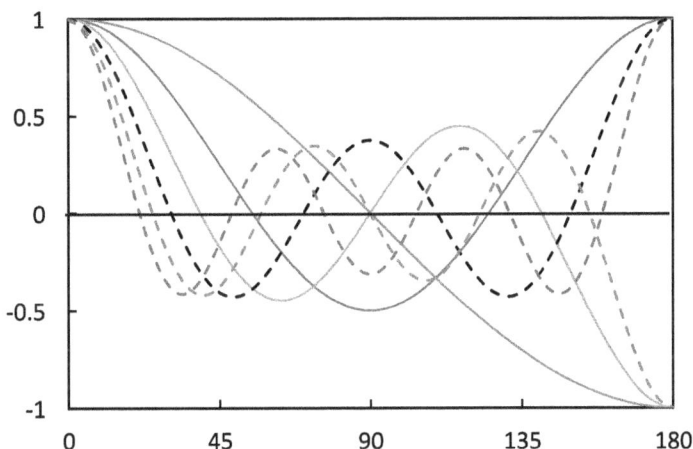

Figure 3.7. Legendre polynomials $P_l(\cos\theta)$, $0 < \theta < \pi$ for $l = 1$ (blue solid), $l = 2$ (red solid), $l = 3$ (black solid), $l = 4$ (blue dashed), $l = 5$ (red dashed), $l = 6$ (black dashed).

$[q_{lm}]$ = length^l × charge. In general, the multipole moments are complex valued; the charges are real so that $q_{lm} = (-1)^m q_{lm}^*$ (where $z^* = x - iy$ is the complex conjugate of the complex number $z = x + iy$).

The term $l = 0$ is the monopole term and the coefficient

$$q_{00} = \sum_i q_i \tag{3.42}$$

is the total charge on the molecule. The radial dependence of the monopole term is $\Phi \sim 1/r$. The terms with $l = 1$ are given by

$$q_{11} = -\sqrt{\frac{3}{8\pi}} \sum_i (x'_i - iy'_i)q_i = -\sqrt{\frac{3}{8\pi}}(p_x - ip_y), \tag{3.43}$$

$$q_{10} = \sqrt{\frac{3}{4\pi}} \sum_i z'_i q_i = \sqrt{\frac{3}{4\pi}} p_z, \tag{3.44}$$

where we define the dipole moment

$$p_x = \sum_i x'_i q_i, \quad p_y = \sum_i y'_i q_i, \quad p_z = \sum_i z'_i q_i. \tag{3.45}$$

The terms with $l = 2$ are

$$q_{22} = \frac{1}{4}\sqrt{\frac{15}{2\pi}} \sum_i (x'_i - iy'_i)^2 q_i = \frac{1}{12}\sqrt{\frac{15}{2\pi}}(Q_{xx} - 2iQ_{xy} - Q_{yy}), \tag{3.46}$$

$$q_{21} = -\sqrt{\frac{15}{8\pi}} \sum_i z'_i(x'_i - iy'_i)q_i = -\frac{1}{3}\sqrt{\frac{15}{8\pi}}(Q_{xz} - iQ_{yz}), \tag{3.47}$$

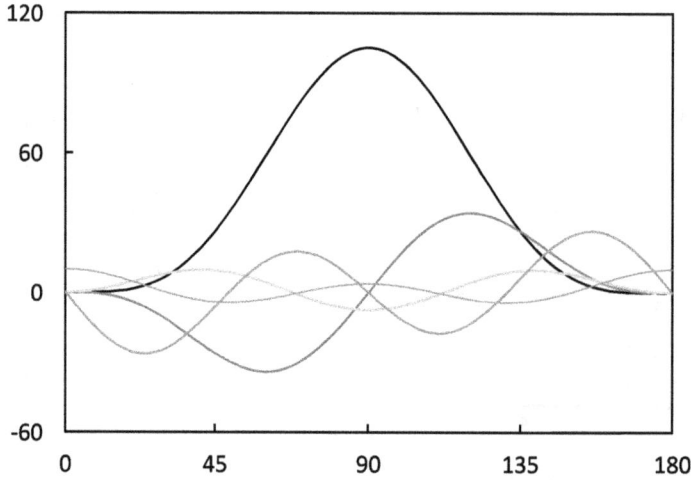

Figure 3.8. The associated Legendre polynomials $P_l^m(\cos\theta)$, $0 < \theta < \pi$ for $l = 4$ and $m = 4$ (black), $m = 3$ (red), $m = 2$ (blue), $m = 1$ (green), $m = 0$ (orange); the values for $m = 1$ and $m = 0$ are multiplied by 10 for better visibility.

$$q_{20} = \frac{1}{2}\sqrt{\frac{5}{4\pi}} \sum_i \left[3(z')_i^2 - (r')_i^2 \right] q_i = \frac{1}{2}\sqrt{\frac{5}{4\pi}} Q_{zz}, \qquad (3.48)$$

where we define the traceless quadrupole moment,

$$Q_{xx} = \sum_i \left[3(x'_i)^2 - (r'_i)^2 \right] q_i, \quad Q_{xy} = Q_{yx} = \sum_i 3x'_i y'_i q_i, \qquad (3.49)$$

$$Q_{xz} = Q_{zx} = \sum_i 3x'_i z'_i q_i, \quad Q_{yz} = Q_{zy} = \sum_i 3y'_i z'_i q_i, \qquad (3.50)$$

$$Q_{yy} = \sum_i \left[3(y'_i)^2 - 3(r'_i)^2 \right] q_i, \quad Q_{zz} = \sum_i \left[3(z'_i)^2 - (r'_i)^2 \right] q_i. \qquad (3.51)$$

The quadrupole moment tensor is defined analogously to the moment of inertia in classical mechanics [3]. It can be shown that the first non-vanishing multipole moment is independent of the choice of the coordinate system. For a neutral molecule, the first two non-vanishing terms can be written in Cartesian coordinates:

$$\Phi(x, y, z) = \frac{1}{4\pi\epsilon_0}\left[\frac{p_x x + p_y z + p_z z}{r^3} \right.$$
$$\left. + \frac{Q_{xx}x^2 + 2Q_{xy}xy + 2Q_{xz}xz + Q_{yy}y^2 + 2Q_{yz}yz + Q_{zz}z^2}{r^5} \right]. \qquad (3.52)$$

We discuss examples of electrostatic potential from a dipole and quadrupole moment in chapter 5, when we discuss the electric fields produced by small molecules.

Qualitatively, the multipole expansion tells us that the angular and radial dependence of the electric field are connected. A point charge, that is the monopole term, produces an electric field that is long range $\Phi(r) \sim 1/r$ and has an angular dependence $\Phi(\theta) = $ const. Higher multipole terms decay more rapidly, $\Phi(r) \sim 1/r^{l+1}$, and have 'more' angular variation, $\Phi(\theta) \sim \cos^l \theta$.

3.3.1 Electric dipoles

A dipole corresponds to the case when we have a positive $q_+ = q$ and a negative charge $q_- = -q$ separated by the distance a so that the total charge vanishes, $Q = q_+ + q_- = 0$. This is called an (electric) dipole. The dipole moment is defined by $\vec{p} = q\vec{a}$, where $\vec{a}$ is the displacement from the *negative* to the *positive* charge. The dipole moment has units $[p] = $ Cm. If a dipole is placed in a *uniform* electric field, the potential energy is given by

$$V = -\vec{p} \cdot \vec{E}. \tag{3.53}$$

The positive and negative charges in a dipole each produce a spherically symmetric electric field that partially cancel out each other. We consider the limiting case $a \to 0$, such that the dipole moment is constant, $p = qa \neq 0$ (mathematically, this would imply that the charges become infinitely large): in this limit, arrangement of a positive and negative charge is referred to as 'point dipole'. The electric field follows,

$$\vec{E}(\vec{r}) = \frac{1}{4\pi\epsilon_0} \frac{3(\vec{p} \cdot \hat{e}_r)\hat{e}_r - \vec{p}}{r^3}, \tag{3.54}$$

where $\hat{e}_r = \vec{r}/r$ is the unit vector along the radius vector. If the electric field is produced by a point dipole $\vec{p_1}$ (taken as the origin of the coordinate system) and another point dipole $\vec{p_2}$ is placed at the point $\vec{r}$, the mutual interaction between the two dipoles can be written:

$$V = \frac{1}{4\pi\epsilon_0} \frac{\vec{p_1} \cdot \vec{p_2} - 3(\hat{e}_r \cdot \vec{p_1})(\hat{e}_r \cdot \vec{p_2})}{r^3}. \tag{3.55}$$

We see that potential energy decays with the inverse third power of the distance between the dipole, $V \sim 1/r^3$, rather than the inverse-square law of the Coulomb law between two (point-) charges. The term 'dipole approximation' is frequently used in the literature. In the example above, the dipole moment $p = q\,a$ only depends on the separation a of the positive and negative charge, and is due to the fact that $q_+ = -q_e$ and the total charge is zero. In the general case, the positive and negative charges differ from each other, $|q_+| = \langle q \rangle + \Delta q/2$ and $|q_-| = \langle q \rangle - \Delta q/2$. We consider the case when the centers of the positive and negative charges are separated by the distance a along the x-axis so that $x_+ = x_0 + a$ and $x_- = x_0$, respectively. The dipole moment follows, $p = (\langle q \rangle + \Delta q/2) \cdot (x_0 + a) - (\langle q \rangle - \Delta q/2) \cdot x_0 = \langle q \rangle a + \Delta q \cdot x_0$, and we conclude that the dipole moment depends explicitly on the choice of the coordinate system.

3.4 Scaled units

The electric field, $\vec{E}$, is determined by the force, $\vec{F}$, on a particle with charge q, $|\vec{E}| = |\vec{F}|/q$, and the magnetic field, $\vec{B}$, is determined by the force, $\vec{F}$, on a particle with charge q moving with velocity $\vec{v}$, $|\vec{B}| = |\vec{F}|/(qv \sin \theta)$, where θ is the angle between $\vec{v}$ and $\vec{B}$. Thus, electric and magnetic properties are characterized by quantities measured in units of the electric charge q in addition to the units of mass, energy, and length. Since electric charge is quantized, we use the elementary charge as the scale for charge:

$$e = 1.609 \times 10^{-19} \text{C}. \tag{3.56}$$

We find the electrostatic potential energy of two elementary charges $q_1 = q_2 = e$ separated by the distance $r = \mathcal{L}$, $V = (4\pi\epsilon_0)^{-1}e^2/\mathcal{L} = 2.307 \times 10^{-18}\text{J} = 1390\,\mathcal{E}$. We thus write the electrostatic potential energy of two point charges q_1 and q_2, cf. equation (3.8),

$$V = 1390\frac{q_1 q_2}{r}. \tag{3.57}$$

Likewise, the magnitude of the Coulomb force between two charges follows:

$$F = 1390\frac{q_1 q_2}{r^2}. \tag{3.58}$$

The arguments in section 3.2 give the classical electron radius $r_e = 1390/$ $[(1/1833) \cdot (3 \times 10^5)^2] = 2.8 \times 10^{-5}$, which gives the lower bound of length.

The obvious choice of the scaled units is $p = e\mathcal{L} = 1.602 \times 10^{-29}$ Cm. However, for molecular systems, the debye unit, D, is still commonly used, and we follow this usage in this book. The debye unit for dipoles is based on the unit 'stat-Coulomb' for static charges, 1 stat-C $\simeq 3.336 \times 10^{-10}$ C, so that 1 D $\simeq 3.336 \times 10^{-30}$ Cm. Since $e\mathcal{L} = 16.09 \times 10^{-30}$ Cm $\simeq 4.823$ D, we arrive at the relation

$$1 \text{ D} = 0.207 \tag{3.59}$$

in scaled units. The energy of a dipole in the external field $\vec{E}$ is written

$$V = -2.01 \times 10^{-9}\frac{\vec{p} \cdot \vec{E}}{\text{D} \cdot \text{Vm}^{-1}}, \tag{3.60}$$

and the dipole–dipole interaction

$$V(\vec{r}) = \frac{60.27}{\text{D}^2}\frac{\vec{p_1} \cdot \vec{p_2} - 3(\vec{p_1} \cdot \hat{e}_r)(\vec{p_2} \cdot \hat{e}_r)}{r^3}. \tag{3.61}$$

The electric potential is defined as the potential energy per unit charge, $V = EPE/q$, and the volt unit is defined as $1\text{ V} = 1\text{J C}^{-1}$. The characteristic scale follows:

$$[V] = \frac{\mathcal{E}}{e} \simeq 1.03 \times 10^{-2}\text{ V}. \tag{3.62}$$

The electric field is then measured in units:

$$[E] = \frac{\mathcal{E}}{\mathcal{L}e} = 1.03 \times 10^{8}\text{ NC}^{-1}. \tag{3.63}$$

In the physics literature, the volt unit is often used to define an energy scale by $1\text{ eV} = e \cdot 1\text{ V}$. We find from equation (2.70) the conversion $1\text{ eV} = 96.8\,\mathcal{E} \simeq 100\,\mathcal{E}$.

The force on a moving charge is given by $F = qvB$ so that the characteristic force $\mathcal{E}/\mathcal{L}$, the elementary charge e, and the characteristic speed $\mathcal{L}/\mathcal{T}$ correspond to a magnetic field $B = [\mathcal{E}/\mathcal{L}]\,[\mathcal{L}/\mathcal{T}]^{-1}e^{-1}$, or

$$[B] = \frac{(\mathcal{E}\mathcal{T})}{(e\mathcal{L}^{2})} \simeq 10 \times 10^{5}\text{ T}. \tag{3.64}$$

This is an extremely powerful magnetic field that only exists on certain neutron stars (magnetars). Magnetic fields available in laboratories are of the order of one tesla (1 T). A charge e moving at a speed $\mathcal{L}/\mathcal{T}$ in a magnetic field $B = 1$ T is subject to the force $F = 1.6 \times 10^{-16}$ N, or in scaled units $F \simeq 9.58 \times 10^{-6}$. Electric currents are measured in units,

$$[I] = \frac{e}{\mathcal{T}} = 1.9 \times 10^{-6}\text{ A}. \tag{3.65}$$

We now write the physical laws in scaled units. The Coulomb law for the force between two charges q_1 and q_2 now reads

$$F = 1390\,\frac{q_1 q_2}{r^2}, \tag{3.66}$$

where it is understood that forces are measured in units $\mathcal{F}/\mathcal{L}$, charges in units of the elementary charge e, and the distance r in units $\mathcal{L}$. Likewise for the electrostatic potential,

$$U = 1390\,\frac{q_1 q_2}{r}, \tag{3.67}$$

where the potential energy is measured in unit $\mathcal{E}$. We use conventional units for external electric and magnetic fields, $[E] = \text{V m}^{-1}$ and $[B] = \text{T}$. The force on a charge in an electric field E is written

$$F = 9.63 \times 10^{-9} \cdot q\frac{\mathsf{E}}{\text{Vm}^{-1}}, \tag{3.68}$$

where the force is in units $\mathcal{E}/\mathcal{L}$ and the force on a moving charge in a magnetic field **B**,

$$F = 9.58 \times 10^{-6} \cdot qv\frac{\mathbf{B}}{\mathrm{T}}, \tag{3.69}$$

where the speed of the charge v is in units $\mathcal{L}/\mathcal{T}$. The force on a current-carrying element with length L follows:

$$F = 1.14 \times 10^{-5} IL\frac{\mathbf{B}}{\mathrm{T}}. \tag{3.70}$$

The oscillatory motion of atoms (ions) and electrons are excited via interaction with an electromagnetic wave. A noticeable transfer of energy from the EM wave to the particle is only possible if the frequency of the EM wave is at (or near) resonance with the frequency of the oscillatory motion of the particle. That is, the two frequencies must be equal, $f_{EM} \simeq f_{particle}$. Because the oscillatory frequency for electrons is about three-orders of magnitude higher than that for ions, we conclude that electrons and ions are excited by different parts of the EM spectrum. In fact, a frequency $f \simeq 10^{13}$ Hz corresponds to an EM wave with wavelength $\lambda = (3 \times 10^8 \, \mathrm{ms}^{-1})/(10^{13} \, \mathrm{Hz}) \simeq 3 \times 10^{-3}$ cm, or

$$f_{EM} \simeq 300 \, \mathrm{cm}^{-1} \; (\mathrm{ions}), \tag{3.71}$$

which corresponds to the infrared (IR) part of the EM spectrum. For electrons, the corresponding frequency is about 1000 times higher, or

$$f_{EM} \simeq 100\,000 \, \mathrm{cm}^{-1} \; (\mathrm{electrons}), \tag{3.72}$$

which corresponds to the UV–vis part of the EM spectrum. We emphasize that these frequencies are based on order of magnitude estimates; in the next chapters, we find more accurate estimates of relevant frequency from more detailed calculations.

References

[1] Arfken G B and Weber H J 2001 *Mathematical Methods for Physicists* 5th edn (San Diego, CA: Harcourt Science–Academic Press)

[2] Feynman R P, Leighton R B and Sands M 1963 *The Feynman Lectures on Physics: Volumes I–III* (Reading, MA: Addison-Wesley)

[3] Goldstein H 1980 *Classical Mechanics* 2nd edn (Reading, MA: Addison Wesley)

[4] Griffiths D J 1999 *Introduction to Electrodynamics* 3rd edn (Upper Saddle River, NJ: Prentice Hall)

[5] Jackson J D 1999 *Classical Electrodynamics* 3rd edn (New York: Wiley)

[6] Purcell E M and Morin D J 2013 *Electricity and Magnetism* 3rd edn (New York: Cambridge University Press)

[7] Schey H M 2004 *Div, Grad, Curl, and All That* 4th edn (New York: W. W. Norton and Company)

Electrostatics at the Molecular Level

Ulrich Zürcher

Chapter 4

Properties of atoms

4.1 Hydrogen atom

A hydrogen atom consists of a proton with charge $q_p = +1$ and an electron with charge $q_e = -1$. The Coulomb force between proton and electron is attractive and the potential energy is negative $V(r) = -E_0/r < 0$,

$$V_C(r) = -\frac{1390}{r}, \tag{4.1}$$

where r is the distance (radius) between proton and electron. The Coulomb potential has the same radial dependence $V \sim r^{-1}$ as the gravitational potential energy between, say, sun and planets, and the orbits of electrons in atoms is often compared with the solar system. In particular, the inverse power law does not define a characteristic length and energy scale; rather, the length scale determines the energy scale and vice versa. It follows that the electron in the hydrogen atom can orbit the proton at any distance and its energy can have any (negative) value. The calculation is straightforward: the electron is kept on a circular orbit by the Coulomb attraction so that $1833^{-1} v^2/r = 1390/r^2$. We find the speed $v = 1596/\sqrt{r}$. The kinetic energy follows, $T = 1390/2r$, so that the total energy is given by $E = V/2 = -695/r$. Thus, for an orbit with radius $r = 1$, the period follows, $T_e = 2\pi/1596 \simeq 4.0 \times 10^{-3}$.

The similarity between the orbits of planets and electrons is flawed, however, because in contrast to the motion of planets, an electron would not be stable since a charge with acceleration a emits radiation, $P_{rad} = e^2 a^2/6\pi\epsilon_0 c^3$ or $P_{rad} = -(2/3)(1390/(3 \times 10^5)^3)a^2 = -8.6 \times 10^{-16}a^2$. Since $a = a_c = v^2/r = 2.55 \times 10^{-6}/r^2$, we find $P_{rad} = -5.58 \times 10^{-3}/r^4$, The radiation of energy causes the electron to lose energy: $dE/dt = (695/r^2) \cdot dr/dt$. We set $dE/dt = P$ and arrive at $-5.58 \times 10^{-3}/r^4 = (695/r^2) \cdot dr/dt$ or $dr/dt \simeq -8.0 \times 10^{-6}/r^2$. We use $r^2 dr = d(r^3)/3$ and find the time-dependence of the orbital radius $r(t) = (r_0^3 - 2.4 \times 10^{-5}t)^3$. For $r_0 = 1$, we find the time of radiative loss, $t_{rad} = (2.4 \times 10^{-5})^{-1} \simeq 4.0 \times 10^4$.

doi:10.1088/978-1-64327-186-6ch4

We calculate the number of revolutions of the electron until it collapses into the proton from t_{rad} and the orbital period T_e, $N^* = t_{rad}/T_e = (4 \times 10^4)/(4 \times 10^{-3}) = 10^7$.

The electron does not collapse into the proton, however, and instead remains at a fixed radius that is of the order of unity $r \simeq 1$. It follows that the typical binding energies of electrons in atoms are of the order $E \simeq 10^3$. This is only possible if the attractive Coulomb force between electron and proton is balanced by a repulsive force acting on the electron. If the repulsive force can be derived from a potential with power-law dependence, $V_R(r) \sim 1/r^n$, the strength must be of the order 10^3, and we arrive at the expression $V_R(r) \simeq 1000/r^n$ with $n > 1$. Since electron and proton are 'sub-atomic' particles, one might suspect that the repulsive potential is due to a fundamental force other than the Coulomb and gravitational force. However, the proton is a baryon, while the electron is a lepton so that the 'strong' force does not act between them. The 'weak' force is many orders of magnitude smaller than the electric force [10] and is irrelevant for atoms and molecules. This shows that the repulsive force must be intrinsic to the nature of the electron.

Bohr recognized in 1913 that the stability of the electron orbit in the hydrogen atom (and of course in any other atom) requires new laws of physics which he identified with the wave-like (or quantum-mechanical) behavior of electrons. The wave-like nature implies that the uncertainty of the momentum of an electron is large if the electron is confined to a small space; loosely speaking, the electron moves fast and has a large kinetic energy. We identify the size of the atom with the uncertainty of the electron position so that the 'typical' momentum is given by $p \simeq \hbar/\Delta x = 0.634/\Delta x$. The kinetic energy of the electron follows, $T = p^2/2m \simeq (2m_e)^{-1}(0.634/\Delta x)^2 \simeq (0.2/m_e)(\Delta x)^{-2}$. We recall that the kinetic energy of the confined electron has the desired power-law behavior with $n = 2$ and diverges as $r \to 0$. The strength of the repulsive force is inversely proportional to the electron mass: the wave nature of particle is more apparent, the smaller its mass. The strength of the repulsive potential yields the relation $1000 \simeq 0.2/m_e$. We find the mass of the electron, $m_e \simeq 1/5000$, which is about 1/3 of the actual value $m_e = 1/1836$. Thus, the separation of positive and negative charges on atomistic scales reflects the wave-like nature of electrons.

In general, the motion of the atom is separated into the motion of the center of mass, $M = m_p + m_e$, and the relative motion of a particle with reduced mass μ around the center of mass. In the case of the hydrogen atom, the proton is much heavier than the mass of the electron, $m_p/m_e \simeq 1836$, so that $M \simeq m_p$ and $\mu_e \simeq m_e \simeq 0.005$. We thus arrive at the usual picture in which the electron orbits the stationary proton. The electron is not the only lepton; the muon and tau-particle are second- and third-generation leptons with properties identical to those of electrons but with bigger mass. The mass of the muon is $m_\mu \simeq 1/8.9$ and the mass of the tau-particle is $m_\tau \simeq 1.89$ so that the reduced masses of the muonic and tauonic hydrogen atoms follow $\mu_\mu = 1/9.9 \simeq 0.1$ and $\mu_\tau \simeq 0.65$. The ratios of the reduced masses follow, $\mu_\mu/\mu_e \simeq 208$ and $\mu_\tau/\mu_e \simeq 3500$. Thus, the size of a muonic and tauonic hydrogen atom would be about 1/200 and 1/3500 of an electronic hydrogen atom (i.e. a 'normal' hydrogen atom) and the binding energies would be $E_\mu \simeq 10^5$ and

$E_\tau \simeq 10^6$. We conclude that phenomena associated with electric charges could not be observed with techniques used in chemistry if the electron mass would be similar to those of the other leptons (muon and tau).

In a quantum-mechanical treatment, one writes the electronic wave function as a product of a radial and an angular part $\psi_{nlm}(r, \theta, \phi) = (1/r)u_n(r)Y_{lm}(\theta, \phi)$, and finds $\mathcal{H}u = Eu$, or

$$-367\frac{d^2u}{dr^2} + \left[-\frac{1390}{r} + 367\frac{l(l+1)}{r^2}\right]u = Eu, \tag{4.2}$$

where we used $\hbar^2/2m_e = 367$. Here, we are interested in the radial dependence and identify the parameter λ (not to be confused with a wavelength) in the Hellmann–Feynman theorem with the radius $\lambda = r$. This would correspond to the case when $d^2u/dr^2 = 0$ and is equivalent with the semi-classical case first considered by Bohr in 1913 [3, 6]. We treat the electron as a wave with wavelength $\lambda = h/p$. We further assume that the trajectory of the electron in the (x, y) plane is perpendicular to the angular momentum $\vec{L} = L\hat{z}$, then an average with respect to the orientation of the z-axis must be taken. The condition that the electron is in a stationary state implies that the electron forms a standing wave: $2\pi r = n\lambda$. We insert the expression for λ and recover the well-known result that the stationarity is equivalent with the quantization of the angular momentum: $L_z = n\hbar$. The kinetic energy of the electron follows from the 'centrifugal potential' and is inversely proportional to the square of the radius: $L_z^2/2\,mr^2 = (n\hbar)^2/2\,mr^2$. Thus the semi-classical expressions yield the correct radial dependence and order-of-magnitude of the repulsive potential, although the interpretation of the quantum number is incorrect. We identify the centrifugal potential with the repulsive potential and obtain

$$V_R(r) = \frac{367\,n^2}{r^2}, \tag{4.3}$$

in agreement with our discussion above. We note that $V_R(r)$ has the desired radial dependence and strength. The total potential for the radial motion is the sum of the Coulomb interaction between proton and electron and a repulsive potential that has quantum-mechanical (wave-like) origin: $V(r) = V_R(r) + V_C(r) = 367\,n^2/r^2 - 1390/r$.

Since we assume that the radius of the electron is fixed, the radial velocity of the electron is zero $v_r = dr/dt = 0$, the contribution of the radial velocity to the kinetic energy is zero $T_r = mv_r^2/2 = 0$, and the energy of an electron in an orbit with a fixed radius r from the proton follows:

$$E(r) = \frac{367\,n^2}{r^2} - \frac{1390}{r}. \tag{4.4}$$

In figure 4.1, the total energy $E(r)$ for $n = 1$ and $n = 2$ is shown. While our derivation is not rigorous, we arrived at an expression for a quantum-mechanical system in the 'spirit' of the Hellmann–Feynman theorem: the radius of the electron orbit around

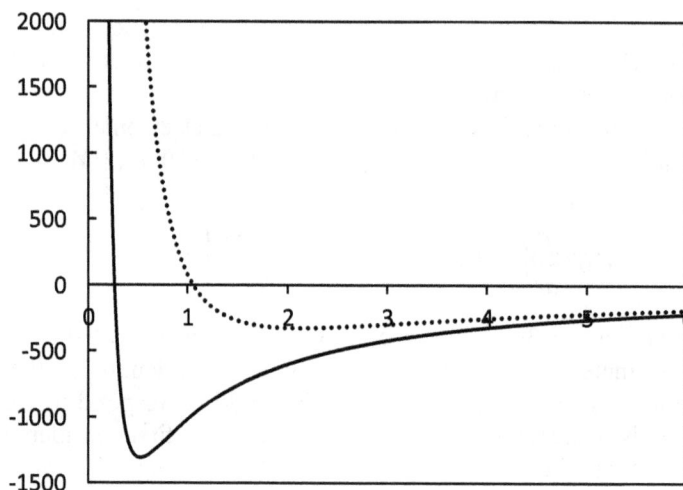

Figure 4.1. The energy of the electron as function of the orbital radius for $n = 1$ (solid) and $n = 2$ (dashed).

the proton is the parameter λ. We set $dE/dr = 0$ and find the equilibrium separation between proton and electron $-2 \cdot 367n^2/r^3 + 1390/r^2 = 0$, or

$$r_n = a_0\, n^2, \tag{4.5}$$

where the Bohr radius is given by

$$a_0 = \frac{2 \cdot 367}{1390} = 0.529. \tag{4.6}$$

The energy can now be written in terms of the radius r_n:

$$E(r) = 1390\left(\frac{r_n}{2r^2} - \frac{1}{r}\right). \tag{4.7}$$

The energy eigenvalue of the electron then follows as the energy at the radius r_n: $E_n = E(r_n)$, or

$$E_n = -\frac{1390}{2r_n}. \tag{4.8}$$

This is typically written as $E_n = -1310/n^2$ with $1310 = 13.6$ eV in the *electronvolts* unit, which is still used in the physics literature.

The electron undergoes uniform circular motion on an orbit with radius r_n. We set the Coulomb force equal to the mass of the electron times the centripetal acceleration $a_c = \omega^2 r_n$: $m\omega^2 r_n = 1390/r_n^2$. We solve for the angular speed $\omega_n = \sqrt{1390/mr_n^3}$ so that

$$\omega_n = \frac{\omega_1}{n^{3/2}}, \tag{4.9}$$

where the angular frequency in the ground states is given by $\omega_1 = \sqrt{1390/ma_0^3}$ so that

$$\omega_1 = \sqrt{\frac{1390}{(1833)^{-1} \cdot (0.529)^3}} = 4150, \tag{4.10}$$

or period $T_n = 2\pi/\omega_n = 1.514 \times 10^{-3}$. In standard SI units, this corresponds to a frequency 2.6×10^{17} Hz or a period $T_n = 2.4 \times 10^{-17}$ s, respectively.

The second derivative of the energy is $d^2E/dr^2 \mid_{r_n} = 1390/r_n^3 > 0$, which implies that the electron orbit with radius r_n is stable under small disturbance $r_n \to r_n + \delta r$. We arrive at a second-order expansion of the electron energy for $\delta r \neq 0$,

$$E(r + \delta r) = E_n + \frac{1}{2}k_n(\delta r)^2, \tag{4.11}$$

where we introduced $k_n = d^2E/dr^2 \mid_{r=r_n}$. We find $k_n = 1390/r_n^3$, or

$$k_n = \frac{9440}{n^6}, \tag{4.12}$$

where we used $k_1 = 1390/a_0^3 = 1390/0.529^3 = 9440$. Equation (4.13) is a harmonic potential for the radial displacement δr. We recall that a quadratic potential $V(\delta r) = k(\delta r)^2/2$ describes a spring force with constant k. This result implies that the combined action of electrostatic attraction and quantum-mechanical repulsion of the electron can be described by an 'effective' spring force with constant k_n, cf. figure 4.2. This spring force is a simple example of a semi-empirical force that enters computational models for (macro-) molecules. This shows that the interplay between electrostatic attraction between electron and proton and quantum-mechanical repulsion leads to an effective proton–electron interaction in the form of a harmonic pseudoforce.

That is, the spring constant is smaller for higher excited states, $k_{n'} < k_n$ for $n' > n$, and the spring becomes 'softer'. If the electron is 'displaced' from the equilibrium δr_0, it undergoes oscillatory motion in a radial direction similar to a block attached to a

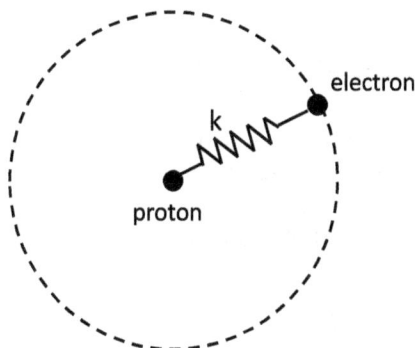

Figure 4.2. Model of the hydrogen atom. The proton is stationary and the orbiting electron is held on its trajectory r_n by a spring with constant k_n.

spring, $\delta r(t) = \delta r_0 \cos(\omega_n t)$, where $\omega_n = \sqrt{k_n/m_e}$ is the (angular) frequency. Here we used the result that the frequency in the radial direction is identical to the orbital frequency, $\omega_n = \sqrt{k_n/m_e}$.

The degeneracy of the orbital and radial frequency implies that the electron orbits in ellipses in non-stationary states $E \neq E_n$. To see this, we assume that the electron is located along the y-axis and the radial motion is at the turning point $\delta r = \delta r_0$ so that $(x = 0, y = r_n + \delta r_0)$ at $t = 0$, it orbits in the (x, y)-plane towards the x-axis and the radial motion returns to the equilibrium $\delta r = 0$ in a quarter period $t = T_n/4 = \pi/2\omega_n$ so that $(x = r_n, y = 0)$. In another quarter period at $t = T_n/2 = \pi/\omega_n$, the electron has orbited towards the $-y$-axis and the radial motion is at the turning point $\delta r = -\delta r_0$ so that $(x = 0, y = -r_n - \delta r_0$; then at three-quarters of a period at $t = 3T_n/4 = 3\pi/2\omega_n$, the electron orbits towards the $-x$-axis and the radial motion returns to the equilibrium $\delta r = 0$ so that $(x = -r_n, y = 0)$. Finally after one period $t = T_n = 2\pi/\omega_n$, the electron returns to the initial position, $(x = 0, y = r_n + \delta r_n)$. We conclude that the orbit of the electron is an ellipse[1].

Our results show that the electron is separated from the proton by an integer-multiple of the Bohr radius at any instant. We find the magnitude of the instantaneous dipole in SI units, $p_H = ea_0 \cdot n^2 = 8.51 \times 10^{-30} \, \text{C m} \cdot n^2$. We thus have the dipole moment of the hydrogen atom in the ground state $p_H = 0.529/(0.207D^{-1})$ or

$$p_H = 2.55 \, \text{D} \cdot n^2. \qquad (4.13)$$

The laws of quantum mechanics (Heisenberg's uncertainty principle) prohibit the notion of a well-specified position of an electron at a time t. We note that the *average* position of the electron coincides with the position of the proton. The hydrogen atom has a rotating dipole moment with vanishing average value $\langle p_H \rangle = 0$.

4.2 Hydrogen atom in static fields

4.2.1 Electric fields

Electric dipoles interact with an external electric field so that the electric energy is given by $V_{\text{int}} = -\vec{p} \cdot \vec{E}$. We find in scaled units,

$$V_{\text{int}} = -2.01 \times 10^{-9} \frac{\vec{p} \cdot \vec{E}}{\text{D} \cdot \text{V m}^{-1}}. \qquad (4.14)$$

Light with intensity 1000 W m^{-2} (approximately equal to the solar constant) corresponds to an electric field of the order $E \simeq 10^3$ V m^{-1}. This dipole moment of the hydrogen atom in the state n thus corresponds to a shift in the energy, $\delta E_n = [2.0 \times 10^{-9} \cdot 2.55 \, \text{D} \, n^2 \cdot 1000 \, \text{V m}^{-1}]/(\text{DV m}^{-1}) \simeq 5 \times 10^{-6} \cdot n^2$, so that the

---

[1] Sommerfeld and others were concerned that Bohr's quantum mechanics required circular and not elliptical orbits as is the case in the purely classical case and extended Bohr's original theory to this more general case. One defines two quantum numbers for the orbital and radial motion $\oint p_\phi r d\phi = n_\phi h$ and $\oint p_r dr = n_r = h$. We write $m = \pm n_\phi$ and $n = n_r + n_\phi$. Circular orbits follow for $m = \pm n$ and elliptical orbits follow for $m = \pm(n - 1)$, $m = \pm(n - 2)$, etc. See Deeney T and Sullivan C 2014 *Am. J. Phys.* **82** 883.

fractional change of the energy would be $\delta E_n/E_n \simeq 4 \times 10^{-9} \, n^4$. Even the electric field in a strong laser light (with intensity $S \sim 10^8 \, \text{W m}^{-2}$) has an electric field of the order $\mathsf{E} \simeq 10^6 \, \text{V m}^{-1}$ would only yield a small perturbation of the energy of the electron, even if it is highly excited (i.e. has a large quantum number n).

While the electron and proton are indeed separated by a finite distance, the electron 'orbits' around the proton and the center of the (time-) average position of the electron coincide with the location of the proton. Thus the average dipole moment is zero $\langle p \rangle = 0$ and there is no change of the electron energy that is linear in the external electric field,

$$\delta E^{(1)} \simeq -\langle p \rangle \mathsf{E} = 0, \tag{4.15}$$

where the superscript refers to 'first order'.

In the presence of a constant external electric field $\vec{\mathsf{E}}$, the proton moves parallel to the electric field while the center of the negative charge moves antiparallel to the electric field. As a result, the centers of the negative and positive charge would be separated so that the force on the charges due to the external electric field is equal to the restoring spring force $e\mathsf{E} = k\delta s$ so that $\delta s = e\mathsf{E}/k$, and the charge separation is smaller for a stiffer spring (higher k). It follows that the hydrogen has an induced dipole moment, $p_{\text{ind}} = e\delta s$. Because the dipole moment is directed from the negative to the positive charge, we find that the induced dipole moment is directed *parallel* to the external electric field. We write

$$\vec{p}_{\text{ind}} = \alpha \vec{\mathsf{E}}, \tag{4.16}$$

which defines the polarizability α of the hydrogen atom. In the semi-classical approximation, the trajectory of the electron is confined to a plane so that we distinguish two cases: (i) the electric field is in the plane of the electron orbit, and (ii) the electric field is perpendicular to the plane of the electron orbit.

(i) We assume that the electric field is along the x-axis, $\vec{\mathsf{E}} = \mathsf{E}\hat{e}_x$, as shown in figure 4.3. We use a coordinate system centered at the new position of the proton, because the electric field along the x-direction corresponds to an electrostatic potential $V(x) = -\mathsf{E}x$, where we set $V(x = 0) = 0$. Because the deviation from the unperturbed trajectory is small, we use the second-order expansion of the energy and arrive at the expression

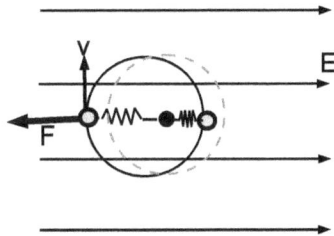

Figure 4.3. Hydrogen atom in a uniform electric field in the plane of the electron orbit.

$$E(r, \phi) = -E_n + \frac{1}{2}k_n(r - r_n)^2 + 9.6 \times 10^{-9} \cdot \frac{\mathsf{E}}{\mathsf{V}\,\mathsf{m}^{-1}} \cdot 2r \cos \phi, \quad (4.17)$$

where we used polar coordinates so that $x = r \cos \phi$ and $y = r \sin \phi$ and ϕ is the angle with respect to the x-axis. We set $dE/dr = 0$ and find the radius

$$r = r_n + 2 \cdot \frac{9.6 \times 10^{-9}}{k_n} \frac{\mathsf{E}}{\mathsf{V}\,\mathsf{m}^{-1}} \cdot \cos \phi, \quad (4.18)$$

or correct in the first order in the external electric field E,

$$r = \frac{r_n}{1 + e \cos \phi}, \quad (4.19)$$

where we introduced the eccentricity $e = 2 \cdot (9.6 \times 10^{-9}/k_n r_n) \cdot \mathsf{E}\mathsf{V}^{-1}\,\mathsf{m}$. We note that equation (4.20) is the equation of an ellipse[2]. The quantity e is then the eccentricity of the ellipse. We have $k_n r_n = (k_1/n^6) \cdot a_0 n^2 = (k_1 a_0)/n^4$ and $k_1 a_0 = (1390/a_0^3)a_0 = 1390/0.529^2 = 4967$ so that $k_n r_n = 4967/n^4$. The eccentricity follows:

$$e_n = 3.9 \times 10^{-12}\, n^4 \cdot \frac{\mathsf{E}}{\mathsf{V}\,\mathsf{m}^{-1}}. \quad (4.20)$$

Since $\cos \phi$ varies between $+1$ and -1, the separation of the average position of the electron from the proton follows,

$$a = \frac{4 \cdot 9.6 \times 10^{-9}}{1390} r_n^3 \cdot \frac{\mathsf{E}}{\mathsf{V}\,\mathsf{m}^{-1}}, \quad (4.21)$$

where we used $k_n = 1390/r_n^3$. It follows that the hydrogen atom has an induced dipole moment, $p_{\mathrm{ind}} = ea$, or

$$\vec{p}_{\mathrm{ind, \parallel}} = \alpha_{\perp} \frac{\vec{\mathsf{E}}}{\mathsf{V}\,\mathsf{m}^{-1}}, \quad (4.22)$$

where we introduced the polarization

$$\alpha_{\parallel} = 4 \cdot \frac{9.6 \times 10^{-9}}{1390} r_n^3. \quad (4.23)$$

(ii) We assume that the electron orbits the proton in the (xy)-plane and apply an electric field along the $+z$-axis, as shown in figure 4.4. The force on the proton (electron) points along the $+z$- $(-z$-$)$ axis. The net force on the hydrogen atom is zero, $\vec{F} = \vec{F}_p + \vec{F}_e$, so that the center of mass is unchanged. We simplify the system and identify the proton with the center-of-mass of the atom. Thus, the proton is fixed, and we only consider

---

[2] The equation of an ellipse is often written in the form $r(\theta) = a(1 - e^2)/(1 - e \cos \theta)$. We easily recover this form by writing $\theta = 180 - \phi$.

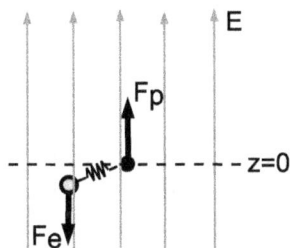

Figure 4.4. Hydrogen atom in a uniform electric field perpendicular to the plane of the electron trajectory.

the displacement of the electron from the (xy)-plane. We note that the centrifugal potential only depends on the perpendicular distance r from the axis of rotation, whereas the Coulomb interaction depends on the separation $\sqrt{r^2 + z^2}$. Since a uniform electric field along the $+z$-axis corresponds to an electric potential $V(z) = -\mathsf{E}z$ (we set $V(z = 0) = 0$):

$$E(r, z) = 1390\left(\frac{r_n}{2r^2} - \frac{1}{\sqrt{r^2 + (2z)^2}}\right) + 9.6 \times 10^{-9} \cdot \frac{\mathsf{E}}{\mathrm{V\ m^{-1}}} \cdot z. \quad (4.24)$$

We set $r = r_n$ and find using $1/\sqrt{r_n^2 + (2z)^2} \simeq 1/r_n - 2z^2/r_n^3$,

$$E_{r_n, z} = -\frac{1390}{2r_n}\left(1 - \frac{4z^2}{r_n^2}\right) + 9.6 \times 10^{-9} \cdot \frac{\mathsf{E}}{\mathrm{V\ m^{-1}}} \cdot 2z. \quad (4.25)$$

We take the derivative with respect to z and arrive at the condition for mechanical equilibrium,

$$\left.\frac{\partial E}{\partial z}\right| = 4k_n z + 2 \cdot 9.6 \times 10^{-9}\frac{\mathsf{E}}{\mathrm{V\ m^{-1}}}, \quad (4.26)$$

where we used $k_n = 1390/r_n^3$. We thus find the equilibrium coordinate of the electron:

$$z_{\mathrm{eq}} = -\frac{9.6 \times 10^{-9}}{2k_n} \cdot \frac{\mathsf{E}}{\mathrm{V\ m^{-1}}}. \quad (4.27)$$

We note that the negative sign implies that the electron is displaced antiparallel to the electric field. Thus the induced dipole moment of the hydrogen atom follows,

$$\vec{p}_{\mathrm{ind,\ \perp}} = \alpha_\perp\frac{\vec{\mathsf{E}}}{\mathrm{V\ m^{-1}}}, \quad (4.28)$$

where

$$\alpha_\perp = \frac{9.6 \times 10^{-9}}{2E_0}r_n^3. \quad (4.29)$$

4-9

Comparison with equation (4.23) yields the relation between the polarizabilities $\alpha_\perp = \alpha_\parallel/8$.

The notion of an electron orbit is borrowed from 'old quantum mechanics' but is irrelevant in a quantum-mechanical treatment of the hydrogen atom. Our results for the 'in' and 'out-of' the plane of the electron are identical except for a numerical factor. We introduce a dimensionless factor $\hat{\alpha}$ of the order unity $\mathcal{O}(\hat{\alpha}) = 1$ and write the polarizability in the ground state $n = 1$:

$$\alpha_H = \hat{\alpha} \cdot \frac{9.6 \times 10^{-9}}{1390} a_0^3 = \hat{\alpha} \cdot 4.95 \times 10^{-12}; \tag{4.30}$$

the factor $\hat{\alpha}$ reflects properties of the electronic structure of an atom and must be calculated by solving an appropriate Schrödinger equation [see 2]. For a 'typical' electric field $\mathsf{E} = 1000 \text{ V m}^{-1}$, the induced dipole moment follows,

$$p_{\text{ind}} \sim 10^{-9} \text{ D}, \tag{4.31}$$

where we used the definition of Debye, cf. equation (3.60). Comparison with equation (4.13) shows that the induced dipole moment is about *nine* orders of magnitude smaller than the *instantaneous* dipole moment. The very small value of p_{ind} reflect the stiffness of the electron orbit and thus the size of the hydrogen atom.

We have seen that this stiffness can be described by a pseudoforce acting on the electron that can be modeled by a linear restoring force. The corresponding spring constant depends on the electronic state $k = k_n$, cf. equation (4.13). If the electron orbit deviates from the stationary orbit $\delta r \neq 0$, the spring is stretched or compressed for $\delta r > 0$ and $\delta r < 0$, respectively, and the electron energy is higher than the value E_n for the state n, $E = E_n + \delta E_n$ with $\delta E_n = k_n(\delta r)^2/2$. Since the induced dipole moment is equal to $p_{\text{ind}} = $ charge $\times$ radial displacement, we write the radial displacement $\delta r = \hat{\alpha} (r_n^3/E_0) \cdot 9.6 \times 10^{-9} \cdot [\mathsf{E} (\text{V}^{-1}\text{ m})]$. It follows that the change in the electron energy is quadratic in the external electric field; we say that the change in the energy δE_n is 'second order':

$$\delta E_n^{(2)} = \frac{1}{2}\frac{E_0}{r_n^3} \cdot \left[\hat{\alpha} \frac{r_n^3}{E_0} \cdot 9.6 \times 10^{-9} \cdot \frac{\mathsf{E}}{(\text{V m}^{-1})}\right]^2 = \frac{\hat{\alpha}^2 r_n^3}{2 E_0} \cdot 9.2 \times 10^{-17} \cdot \left(\frac{\mathsf{E}}{\text{V m}^{-1}}\right)^2. \tag{4.32}$$

For the ground state $n = 1$, we find (we use $\hat{\alpha} = 1$)

$$\delta E_1^{(2)} \simeq 5 \times 10^{-21}\left(\frac{\mathsf{E}}{\text{V m}^{-1}}\right)^2. \tag{4.33}$$

The change in the energy is quadratic in the external electric field; this is called the *quadratic Stark effect* [5]. Thus, for a 'typical' macroscopic electric field $\mathsf{E} = 1000 \text{ V m}^{-1}$, we find $\delta E_1 \simeq 5 \times 10^{-15}$, which is a very small change and can generally be ignored.

4.2.2 Magnetic fields

In this subsection, we use e for the charge of the electron for clarity ($e = 1$ in scaled units). Although it is outside the realm of electrostatics, we discuss the response to an external magnetic field. The electron orbits with period τ_n on a circle with radius r_n so that it corresponds to a loop with current $I = e/\tau_n$ and area $A = \pi r_n^2$. Thus the electron has a magnetic (dipole) moment $\mu = IA = (e/\tau_n)(\pi r_n^2) = (e/2m) \cdot m\,(2\pi r_n/\tau_n)r_n = (e/2m)L$, where $L = mv_n r_n$ is the angular momentum. The ratio $e/2m$ is known as gyromagnetic ratio.

In figure 4.5 we draw the situation where the electron orbits the proton *clockwise*. Because the electron is negatively charged, this corresponds to a *counterclockwise* current so that the magnetic moment vector would be directed *out-of-the page*, which we refer to as the z-axis (according to the so-called 'second-right-hand rule'). The external magnetic field is also directed out-of-the-page so that $\vec{B} = B_0\,\hat{e}_z$. In general, the orbital plane of the electron does not have to be perpendicular to the external magnetic field. The orbital plane can be tilted so that the magnetic moment $\vec{\mu}$ makes the angle θ with respect to the external magnetic field. The corresponding energy is given by

$$\delta E^{(1)} = -\vec{\mu} \cdot \vec{B} = -\mu B_0 \cos\theta. \tag{4.34}$$

The angular momentum of the electron is quantized, $L = \hbar L'$, where L is the 'length' of the angular momentum. We find in scaled units

$$\delta E^{(1)} = -5.55 \times 10^{-3} L \cos\theta\, \frac{B_0}{T}. \tag{4.35}$$

This expression is referred to as the *linear* Zeeman term. We note that $L\cos\theta$ is the z-component of the angular momentum for which usually the quantum number m (for 'magnetic', not to be confused with the electron mass) so that $\delta E = -5.55 \times 10^{-3} m(B_0/T)$, because the interaction is negative $\delta E < 0$ when the magnetic moment associated with the orbiting electron is aligned with the external magnetic field. If we consider a sample of many hydrogen atoms, this alignment will yield a macroscopic magnetization $\vec{M}$ is the same direction as the external magnetic field: we write $\vec{M} = \chi\vec{B}$, which defines the magnetic susceptibility. We have $\chi > 0$,

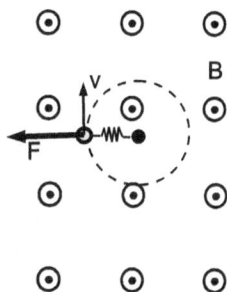

Figure 4.5. Hydrogen atom in a uniform magnetic field perpendicular to the plane of the electron orbit.

which implies that the linear Zeeman term contributes to the *para*-magnetic response of the orbiting electron.

We have only considered the orientation of the plane of the electron so far, but have ignored any change in the radius of the electron orbit. We assume that the magnetic field points out-of-the-page and the electron travels clockwise. The Lorentz force points in the radial direction outwards. The direction of the force is inwards, if the direction of the magnetic field is kept fixed and the direction of the electron orbit is reversed so that it travels *counterclockwise*. The magnitude of the force is $F_B = ev_nB$. This force is balanced by the spring force $ev_nB = k_n\delta r_n$ where δr_n is the deviation $r_n \rightarrow r_n + \delta r$. We find $\delta r = (e\sqrt{E_0/mr_n}B)/(E_0/r_n^3)$, or

$$\delta r = e\sqrt{\frac{r_n^5}{mE_0}}\,B. \tag{4.36}$$

We then find the change in the energy from the expression for the spring ('elastic energy') $\delta E_2 = k_n(\delta r)^2$. Since $k_n = E_0/r_n^3$, we find

$$\delta E^{(2)} = \frac{1}{2}\frac{e^2r_n^2}{m}\cdot B^2. \tag{4.37}$$

The treatment obscures whether the term quadratic in the external magnetic field contributes to the *dia*- or *para*-magnetic response of the hydrogen atom. To see this, we start from the equation of motion of the electron around the proton. Since $v = \omega r$, we obtain $m\omega^2 r = (E_0/r^2) - e\omega rB$. We set $r = r_n$ and use $\omega_n = 2\pi/\tau_n = E_0/(mr_n^3)$ to find $\omega^2 = \omega_n^2 - 2\omega_L\omega$. Here, we introduced the (classical) *Larmor* frequency:

$$\omega_L = \frac{eB}{2m} \tag{4.38}$$

(recall that $\omega_c = eB/m$ is the cyclotron frequency and $\omega_c = 2\omega_L$). We write $\omega \rightarrow \omega_n + \delta\omega$ so that $\omega^2 \rightarrow \omega_n^2 - 2\omega_n\delta\omega$ and find the frequency change $\delta\omega = -2\omega_L$. It follows that the electron orbits the proton at a lower frequency (or longer period) and the magnitude of the current is reduced $I \rightarrow I + \delta I$ with

$$\delta I = -e\frac{\omega_L}{2\pi} = -\frac{e^2B}{4\pi m}. \tag{4.39}$$

The current $I_n = e\omega_n$ produces a (non-uniform) magnetic field $\vec{B}_n$ directed out-of-the-page so that the corresponding flux through the area enclosed by its orbit is positive $\Phi_n > 0$. The external magnetic field in figure 4.5 will produce a *positive* flux $\Phi_{\text{ext}} > 0$. According to Lenz' law, the induced EMF will induce a current so as to oppose the increase in the flux due to the external magnetic field. We thus see that the second-order term $\delta E^{(2)}$, cf. equation (4.37), is associated with the *diamagnetic* response of the hydrogen atom. We used classical mechanics so that our results are only approximation; however, a full quantum-mechanical treatment confirms the dia-magnetic response [5].

Since $\pi r^2 = \pi(x^2 + y^2)$ we find the change of the magnetic moment $\delta\mu = \mu_0 \delta I \pi r^2 = -e^2 B r^2/4m$. We assume spherical distribution of the electron in the hydrogen atom and thus have $\langle x^2 \rangle = \langle y^2 \rangle = \langle z^2 \rangle$ so that in particular $r^2 \rightarrow (2/3)\langle \vec{r}^2 \rangle$. We thus arrive at the Langevin equation for the magnetic moment:

$$\delta\mu = -\mu_0 \frac{e^2 \langle \vec{r}^2 \rangle}{6m} B. \tag{4.40}$$

Here the negative sign ensures that the term is magnetization is opposite to the external magnetic field: we say that the hydrogen atom is *diamagnetic*.

4.3 Electronic transition

The orbit of the electron in a hydrogen atom has properties that are reminiscent of the elastic properties of spring and found the corresponding spring constant k_n for each state n. However, this approximation is only valid for deviation of the radius of the electron orbit from the equilibrium value $\delta r = r - r_n$ that is small compared to r_n, i.e. for $\delta r/r_n \ll 1$. In particular, it cannot be used to describe the transition of the electron from the state n to the state $n' = n \pm 1$.

The wave nature of electrons implies that their orbit is discrete with radii $r_n = 0.529\, n^2$ with frequency ω_n and energies $E_n = -1310/n^2$. $n = 1, 2, 3, \ldots$. Thus, the transition from the ground to the first-excited state is associated with a change in the energy:

$$\Delta E_{21} = -1310\left(\frac{1}{2^2} - 1\right) \simeq 1000. \tag{4.41}$$

We compare this energy to the energy of the hydrogen in a 'typical' magnitude of the electric field 10^3 V m^{-1}, the dipole energy follows from equation (3.60):

$$V_{\text{dipole}} = 2 \times 10^{-9} \frac{2.55\, \text{D} \cdot 10^3\, \text{V m}^{-1}}{\text{D} \cdot \text{V m}^{-1}} \simeq 5 \times 10^{-6}. \tag{4.42}$$

We find the ratio

$$\frac{\Delta E_{12}}{V_{\text{dipole}}} \simeq \frac{1 \times 10^3}{5 \times 10^{-6}} = 2 \times 10^8. \tag{4.43}$$

If we were to treat the transition as resonance phenomena, our calculation would imply that the electron 'acquires' the energy ΔE_{12} in $N \simeq 10^8$ small 'chunks' V_{dipole} until it is in the excited state. We estimate the average frequency from the orbital frequencies in the ground and excited states, $\langle \omega \rangle = (\omega_1 + \omega_2)/2 = (4150 + 4150/8)/2 \simeq 2000$, so that we arrive at an estimate for the transition time, $t_{12} = N \cdot 2\pi/\langle \omega \rangle \simeq 2 \times 10^8 \cdot (2\pi/2000) = 10^5$, or $t_{12} = 10^{-8}$ s $= 10$ ns.

However, the electron is not 'allowed' to orbit the proton at a radius $r_1 < r < r_2$, which is referred to as 'spatial quantization' in the old quantum theory. Rather, the electron is either in the state $n = 1$ or $n = 2$ and the transition must occur instantaneously. More precisely, the time when the transition is uncertain $\Delta t > 0$ so that the uncertainty principle implies that the electron energy is uncertain as well, $\Delta E \simeq \hbar/2\Delta t$. It follows that the electronic state is described as a superposition of the initial and final state, $\psi(\vec{r}, t) = A_1\psi_1(\vec{r})\exp(iE_1t/\hbar) + A_2\psi_2(\vec{r})\exp(iE_2t/\hbar)$.

A classical analog of this state are two strings vibrating with frequencies ω and ω' and secured on the same two boards. This allows the strings to exchange energy. The two strings vibrate 'in phase' and 'out-of-phase' so that the interference of sound waves changes from constructively to destructively, and then back to constructively, and so on. While the resulting sound wave oscillates with a frequency equal to average frequency $(\omega + \omega')/2$, the amplitude varies periodically with a frequency equal to the difference $\omega - \omega'$. This implies that the sound changes from loud to quiet and back to loud and so on with a period $\tau = 2\pi/|\omega - \omega'|$; this variation of loudness is referred to as 'beat'.

The two strings correspond to the electronic states (e.g. the ground and excited states) and the board corresponds to the electric field. This analogy suggests that an electronic transition requires the presence of an electric field that oscillates with frequency $\Delta\omega = \Delta E_{21}/\hbar$,

$$\vec{E}(t) = \vec{E}_0 \sin(\Delta\omega)t = \vec{E}_0[\sin(E_1t/\hbar) \cdot \cos(E_2t/\hbar) - \cos(E_1t/\hbar) \cdot \sin(E_2t/\hbar)], \quad (4.44)$$

where we used an addition theorem of trigonometric functions $\sin(x - y) = \sin x \cos y - \cos x \sin y$. Because the transition requires a 'transfer' of large energy from the electromagnetic field to the electron in an instant, the electromagnetic field must be treated within a particle-like picture, similar to the photoeffect. That is, the transition is described in terms of an emission or absorption of a photon with energy $E_\gamma = \hbar\Delta\omega = \Delta E_{12}$. We generalize to arbitrary initial and final states, and observe that the condition for the frequency of the electromagnetic field has the form of the conservation of energy. When the electron makes a transition from the state n with energy E_n to the state n' with energy $E_{n'}$, the difference is carried away or supplied by a photon with energy E_γ:

$$E_n \pm E_\gamma = E_{n'}. \quad (4.45)$$

For a hydrogen atom, the transition $n = 2 \longrightarrow n' = 3$ (in the Balmer sequence) requires a photon energy $E_\gamma = 1310(1/2^2 - 1/3^2) = 145$; this corresponds to a photon with frequency, cf. equation (2.57),

$$f = \frac{145}{1.2 \times 10^{-2}}\text{cm}^{-1} \simeq 12\ 000\ \text{cm}^{-1}, \quad (4.46)$$

and thus belongs in the visible part of the electromagnetic spectrum, cf. table 2.3.

4.4 Many-electron atoms

While equations (4.5) and (4.8) are the correct expressions for the energy eigenvalues and radius of the electron, this interpretation of the quantum number n is, however, incorrect. In the derivation, n is associated with the angular momentum along the z-direction when in fact it is the so-called principal quantum number n. The quantum number associated with the orbital angular momentum is usually labeled l so that $|\vec{L}| = \hbar\sqrt{l(l+1)}$ and has an upper bound $l \leqslant n - 1$. The z-component of the angular momentum is also quantized and one finds $L_z = m\hbar$ with $m = -l, -l+1, \ldots, -1, 0, 1, \ldots, l-1, l$ (here m must not be confused with the mass m_e of the electron). The eigenvalues of the electron in the hydrogen atom are independent of l and m, and we can say that the magnitude and direction of the angular momentum of the electron are not known. Terminology from atomic spectroscopy is used to label the state l: $l = 0$—s-state, $l = 1$—p-state, $l = 2$—d-state, and $l = 3$—f-state.

The solution of the Schrödinger equation of the electron in the hydrogen atom is written in spherical coordinates (r, θ, ϕ) and the wave equation is written as a product, $\psi(\vec{r}) = R(r)\Theta(\theta)\Phi(\phi)$. One finds that the angular part is a spherical harmonic $\Theta(\theta)\Phi(\phi) = Y_{lm}(\theta, \phi)$, where Y_{lm} are the spherical harmonics familiar from the multipole expansion of the electric field. The notation $Y_{lm}(\theta, \phi) = Y_l^m(\theta, \phi)$ is often used. For $l = 1$, the spherical harmonic $Y_l^m(\theta, \phi)$ is real and aligned with the z-axis, and the notation $p_z(\theta, \phi) = Y_1^0(\theta, \phi) = \cos\theta$ is used. The spherical harmonics $Y_l^\pm(\theta, \phi)$ are complex-valued and often the linear combinations $p_x(\theta, \phi) = (1/\sqrt{2})[Y_1^1(\theta, \phi) + Y_1^{-1}(\theta, \phi)] = \sin\theta\cos\phi$ and $p_y(\theta, \phi) = (1/\sqrt{2})[Y_1^1(\theta, \phi) - Y_1^{-1}(\theta, \phi)] = \sin\theta\sin\phi$ are introduced, and are commonly referred to as 'orbitals'. This method is not restricted to $l = 1$ and can also be used for higher orbitals. For example, for $l = 2$, one finds for $m = 0$: $d_{z^2}(\theta, \phi) = (3\cos^2\theta - 1)$, for $m = \pm 1$: $d_{xz}(\theta, \phi) = \sin\theta\cos\theta\cos\phi$ and $d_{yz}(\theta, \phi) = \sin\theta\cos\theta\sin\phi$, and for $m = \pm 2$: $d_{x^2-y^2}(\theta, \phi) = \sin^2\theta\cos 2\phi$ and $d_{xy} = \sin^2\theta\sin 2\phi$.

The density $|\psi(\vec{r})|^2$ is often referred to as the 'electron cloud'. The charge distribution associated with an electron state follows, $\rho(r, \theta, \phi) = e|\psi(r, \theta, \phi)|^2$. For an electron in an s-state, the density distribution is spherically symmetric and only depends on the distance from the nucleus r, $\rho = \rho(r)$, while in states with higher angular momentum, the charge density also depends on the angles $\rho = \rho(r, \theta, \phi)$. This shows that the electrostatic properties of atoms is determined by a positive charge $Q_+ = n_+e$ at the center and distribution of negative charge described by a density $\rho_-(r, \theta, \phi)$ such that

$$\int_0^\infty dr\, r \int_0^\pi d\theta \sin\theta \int_0^{2\pi} d\phi \rho(r, \theta, \phi) = -n_- \cdot e. \tag{4.47}$$

The net charge on the atom then follows as $Q = (n_+ - n_-)e$.

A neutral atom has the same number of protons and electrons and the net charge is zero, $Q = 0$. We consider a neutral atom with atomic number Z so that the charge of the nucleus is $Q_+ = +Ze$ and Z electrons. In a simple model, the nucleus is surrounded by a spherical 'electron cloud' with charge $Q_- = -Ze$. The (average) charge can be estimated from the charge $-Ze$ and volume of a sphere with radius R, $V = (4\pi/3)R^3$,

$$\rho_- = -\frac{3}{4\pi}\frac{Z}{R^3}. \tag{4.48}$$

For example, for the hydrogen atom, we find

$$\rho_{-, \text{H}} = \frac{3 \cdot 1}{4\pi \cdot 0.529^3} = 1.61, \tag{4.49}$$

or $\rho_{-, \text{H}} = (3/4\pi) \cdot (1.609 \times 10^{-19}\ \text{C})/(0.529 \times 10^{-10}\ \text{m})^3 = 6.81 \times 10^{10}\ \text{C m}^{-3}$. We list in table 4.1 the first ionization energy E_0 (corresponding to the least bound electron) and the average density of the electron cloud ρ_- for the first 20 elements in the periodic table. In figure 4.6, we show the ionization energy E_0 and ρ_- vs the atomic (or proton) number Z.

4.4.1 Atom in a uniform electric field

We now examine the effect of an uniform electric field $\overrightarrow{E}$ on the atom. We follow the derivation in [9] and assume that the electron cloud is not distorted. In figure 4.7 we apply a uniform electric field $\overrightarrow{E}$ along the horizontal and directed towards the right. We take the x-axis along the horizontal and identify $x = 0$ with the (identical) center of the atom in the zero electric field. The force on the nucleus is directed towards the right, $\overrightarrow{F}_{E,\,+} = Z\overrightarrow{E}$, and the force on the center of the electron cloud is directed towards the left, $\overrightarrow{F}_{E,\,-} = -Z\overrightarrow{E}$ (recall that the charge is measured in units of e),

$$|F_{\text{E}}| = 9.63 \times 10^{-9} Z \cdot \frac{\text{E}}{\text{V m}^{-1}}. \tag{4.50}$$

The effect of these two forces would tear apart the electron. However, this does not happen, and the centers of the positive and negative charges are displaced by some finite distance $x_0 < \infty$. This requires a restoring force that increases with the separation between positive charge and the center of negative charge distribution. Clearly, this force is due to Coulomb attraction between the positively charged center of the atom and the negatively charged electron cloud. This Coulomb attraction on the positive and negative charge is directed towards the left and right, respectively, as is required to oppose the force due to external electric field.

Because atoms are stable (that is, the electron cloud does not separate from the nucleus) in the presence of an external electric field, we need to show that the magnitude of the attractive Coulomb force *increases* with increasing separation x. To this end, we assume that the nucleus is at $+ x/2$ and the center of the electron cloud is at $-x/2$. We recall from section 3.2 that the positively charged nucleus is attracted to the fraction of the negatively charged electron cloud inside the sphere of radius x:

$$F_\text{C} = 1390\frac{(Z)\delta Q_-}{x^2}. \tag{4.51}$$

Table 4.1. Properties of atoms. The ionization energy E_0 in units [eV (kJ^{-1} mol)], radius R in units 10^{-10} m, and the charge density of the 'electron cloud' ρ_- in dimensionless units/10^{10} C m^{-3} (data taken from [8], p 378).

Atom (Z)	E_0	R	ρ_-
H (1)	13.505/1307	0.529	1.61/6.18
He (2)	24.580/2379	0.291	9.68/37.2
Li (3)	5.390/522.3	1.586	3.14/12.1
Be (4)	9.320/903	1.040	0.85/3.26
B (5)	8.296/804	0.776	2.55/9.80
C (6)	11.264/1091	0.620	6.01/23.1
N (7)	14.54/1409	0.521	11.82/45.4
O (8)	13.614/1319	0.450	20.96/80.5
F (9)	17.42/1688	0.396	34.60/132.9
Ne (10)	21.559/2089	0.354	53.81/206.6
Na (11)	5.138/498	1.713	0.52/2.00
Mg (12)	7.644/741	1.279	1.37/5.26
Al (13)	5.984/580	1.312	1.37/5.26
Si (14)	8.194/794	1.068	2.74/10.5
P (15)	11.00/1066	0.919	4.61/17.7
S (16)	10.357/1004	0.810	7.19/27.6
Cl (17)	13.01/1261	0.725	10.65/40.9
Ar (18)	15.755/1527	0.659	15.02/57.7
K (19)	4.339/420	2.162	0.45/1.73
Ca (20)	6.111/592	1.690	0.99/3.80

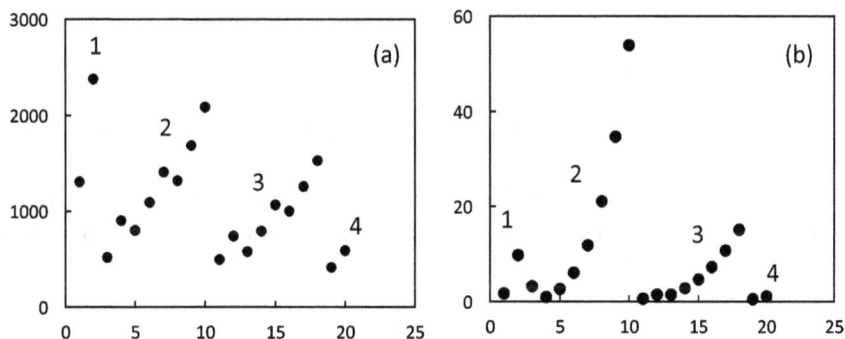

Figure 4.6. (a) The ionization energy E_0 vs the atomic (or proton) number Z; (b) the charge density ρ_- of the electronic cloud vs the atomic (or proton) number Z. The labels (1, 2, 3, and 4) refer to the series of the periodic table.

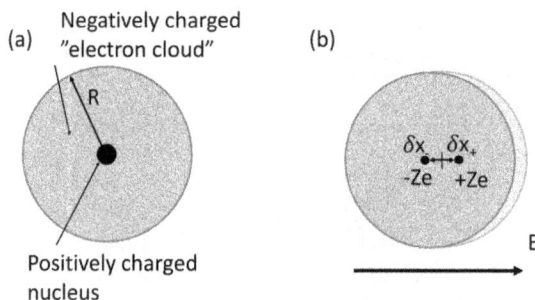

Figure 4.7. (a) A neutral atom when no external electric field is present. The center of the negatively charged electron cloud coincides with the positively charged nucleus. (b) A neutral atom when an external electric field is present. The center of the negatively charged electron cloud coincides and is shifted opposite to the electric field and the positively charged nucleus is shifted along the electric field.

The charge δQ_- is proportional to the volume of a sphere with radius x, cf. figure 4.8,

$$\delta Q_- = \rho_- \frac{4\pi}{3} x^3 = -Z \left(\frac{x}{R} \right)^3. \tag{4.52}$$

The magnitude of the Coulomb force follows,

$$F_C = 1390 \frac{Z \cdot [Z \, (x/R)^3]}{x^2} = 1390 \frac{Z^2}{R^3} x; \tag{4.53}$$

that is, the attractive Coulomb force between the nucleus and the electron cloud is described by a linear restoring force with constant $k = 1390 \, Z^2/R^3$. We note that the constant is of the order $\mathcal{O}(k) \sim 10^3$, which is consistent with our estimate in chapter 2, cf. equation (2.64). This attractive Coulomb force is balanced by the electric force due to the external field; we set $F_E = F_C$ and find $9.63 \times 10^{-9} Z \cdot E$ (V^{-1} m) $= 1390 \, (Z^2/R^3)x$ so that the displacement is proportional to the external field:

$$x = \frac{9.63 \times 10^{-9} \cdot R^3}{1390 \, Z} \frac{E}{V \, m^{-1}} = 6.92 \times 10^{-12} \cdot \frac{R^3}{Z} \frac{E}{V \, m^{-1}}. \tag{4.54}$$

We assume a typical $E \simeq 10^3$ V m^{-1} and find for the hydrogen atom ($Z = 1$ and $R = 0.529$), $x_H = 1.02 \times 10^{-9}$ and for oxygen ($Z = 8$ and $R = 0.45$), $x = 1.10 \times 10^{-10}$; that is, an electric field associated with sunlight barely displaces the electron cloud from the nucleus.

Because the centers of the positive and negative charges are displaced from each other in the presence of an external electric field, the atoms have an *induced* dipole moment given by $p_{ind} = Z \cdot x$, or

$$p_{ind} = 6.92 \times 10^{-12} \, R^3 \cdot \frac{E}{V \, m^{-1}}, \tag{4.55}$$

so that the induced dipole moment is proportional to the volume of the atom (since $R^3 \sim V$) and is independent of the atomic number Z. We assume the $E = 10^3$ V m^{-1}, and find the induced dipole moment for hydrogen $p_{ind, \, H} = 1.02 \times 10^{-9}$ or

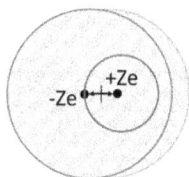

Figure 4.8. The charge inside a radius δr is proportional to the volume $\delta q = Ze(\delta r/R)^3$.

$p_{ind, H} = 4.95 \times 10^{-9}$ D and for oxygen $p_{ind, O} = 6.31 \times 10^{-10}$ or $p_{ind, O} = 3.05 \times 10^{-10}$ D (where we used the conversion for one debye 1 D = 0.207, cf. equation (3.59)). The polarizability is proportional to the electric field $p \sim E$, and the expression for the induced dipole moment defines the polarization $p = \alpha E$. We use scaled units $\alpha = p_{ind}/[9.63 \times 10^{-9}E/(V\,m^{-1})]$ and find

$$\alpha_{approx} = \frac{1}{1390} \cdot R^3. \tag{4.56}$$

We find for the hydrogen atom $\alpha_H = (1/1390)(0.529)^3 = 1.1 \times 10^{-4}$ and for the oxygen atom $\alpha_O = (1/1390)(0.45)^3 = 6.6 \times 10^{-5}$. In SI units, the polarization is given by $\alpha = 4\pi\epsilon_0 R^3$; alternatively, polarizations are often given in terms of 'atomic units' defined as the polarizability of the hydrogen atom,

$$1\,au = 4\pi \cdot 8.85 \times 10^{-12}\,C^2\,N^{-1}\,m^{-2} \cdot (0.529 \times 10^{-10}\,m)^3$$
$$= 1.646 \times 10^{-40}\,C\,m^2\,V^{-1}. \tag{4.57}$$

The conversion between atomic and scaled units follows: 1 au = 1.1×10^{-4}. The approximate expression for the atomic polarizability can be written in atomic units as $\alpha_{approx} = (R/0.529)^3$.

We calculate the electric field E^* that would produce an induced dipole moment equal to one debye: $p_{ind} = 1$ D $= \alpha \cdot E^*$. We find in scaled units

$$E^* = \frac{1\,D}{1\,au} = 1390\frac{0.207}{0.529^3} = 1944, \tag{4.58}$$

or in SI units, $E^* = (3.336 \times 10^{-30}\,C\,m)/(1.646 \times 10^{-40}\,C\,m^2\,V^{-1}) = 2.04 \times 10^{10}$ V m^{-1}. Such an enormous electric field does not exist at macroscopic scales (as it exceeds the electric breakdown value for air $E_{air, max} \simeq 3 \times 10^6$ V m^{-1}). In table 4.2 we list the atomic polarizabilities from a quantum-mechanical calculation from [7] and compare them with the polarizability derived from equation (4.56) (figure 4.9). It is interesting to consider a numerical example. We place a sodium atom in direct sunlight so that is exposed to an electric field of strength $E_{sun} \simeq 10^3$ V m^{-1} or $E \simeq 10^{-5}$ in scaled units. The dipole moment follows, $p = \alpha E$ or $p = 16.5 \cdot 10^{-5}$ D $= 1.65 \times 10^{-4}$ D. Since the charge is $Q = Z = 11$, we find the displacement of the center of the electron cloud and the nucleus $\delta x = p/Q = 1.4 \times 10^{-4}$. The radius of the neutral sodium atom is $R = 1.713$ so that the fractional distortion of the charge distribution follows, $\delta x/R = (1.4 \times 10^{-4})/(1.713) \simeq 8 \times 10^{-6}$.

Table 4.2. Atomic polarizabilities in atomic units: from a quantum-mechanical calculation and from the ratio $\alpha_{num}/\alpha_{approx}$ for $1 \leqslant Z \leqslant 20$.

Atom (Z)	α_{num}	α_{approx}	$\alpha_{num}/\alpha_{approx}$
H (1)	4.50	1	4.5
He (2)	1.38	0.166	8.3
Li (3)	164	26.9	6.1
Be (4)	38.8	7.60	5.1
B (5)	20.5	3.16	6.5
C (6)	11.3	1.61	7.0
N (7)	7.43	0.955	7.8
O (8)	6.00	0.616	9.7
F (9)	3.74	0.419	8.9
Ne (10)	2.67	0.300	8.9
Na (11)	162	34.0	4.8
Mg (12)	71.0	14.1	5.0
Al (13)	58.0	15.3	3.8
Si (14)	37.1	8.23	4.5
P (15)	24.8	5.24	4.7
S (16)	19.5	3.59	5.4
Cl (17)	14.7	2.57	5.7
Ar (18)	11.1	1.93	5.8
K (19)	292	68.2	4.3
Ca (20)	160	32.6	4.9

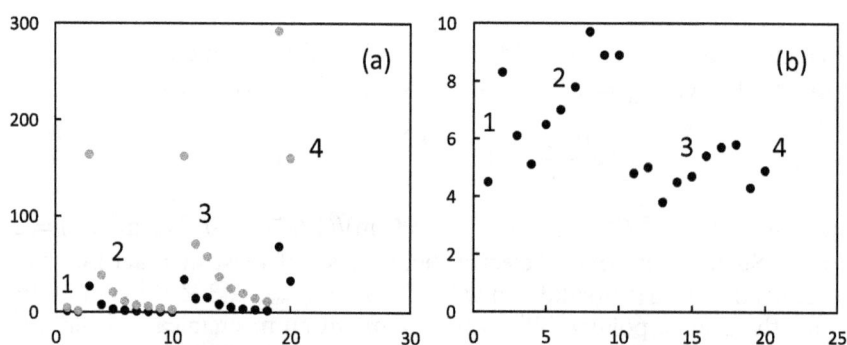

Figure 4.9. (a) Atomic polarizabilities in atomic units for the first 20 elements: numerical value α_{num} (green), the approximate expression equation (4.56) (black) for $1 \leqslant Z \leqslant 20$. (b) The ratio $\alpha_{num}/\alpha_{approx}$ for $1 \leqslant Z \leqslant 20$.

4.4.2 Polarization

For a macroscopic sample, the dipole moment $\vec{p}$ of atoms and molecules yields the polarization $\vec{P}$,

$$\vec{P} = n\vec{p}.$$

(4.59)

In SI units, we have $[n] = $ m^{-3} and $[p] = $ C m so that $[P] = $ Cm^{-2}, where N is the average number of molecule per unit volume. The sum of the electric field $\vec{E}$ and polarization $\vec{P}$ is the electric displacement [4],

$$\vec{D} = \epsilon_0 \vec{E} + \vec{P}. \tag{4.60}$$

Since the polarization is proportional to the electric field, the polarization is written

$$\vec{P} = \epsilon_0 \chi_e \vec{E}, \tag{4.61}$$

where χ_e is the electric susceptibility. The electric displacement is written

$$\vec{D} = \epsilon_0 (1 + \chi_e) \vec{E} = \epsilon_0 \epsilon_r \vec{E}, \tag{4.62}$$

where $\epsilon_r = 1 + \chi_e$ is the relative electric permittivity.

We write the polarization in terms of $\vec{E}$ and ϵ_r: $P = (\epsilon_E - 1)\epsilon_0 E$. We use the definition of polarization and express the dipole moment in terms of the polarization to find $P = np = n\alpha E = (\epsilon_r - 1)\epsilon_0 E$. It follows that $N\alpha = (\epsilon_r - 1)\epsilon_0$. In SI units we have $\alpha = 4\pi\epsilon_0 R^3$ so that we find the expression for the relative electric permittivity:

$$(\epsilon_r - 1) = 4\pi n R^3. \tag{4.63}$$

We get a 'sense' for a typical value if we assume the number density is equal to that of the ideal gas at standard conditions and the size of the atom is equal to the length scale $\mathcal{L}$. Since the volume of one mole of an ideal gas is 22.4 L, we find the number density $n_0 = 6.02 \times 10^{23}/(22.4 \times 10^{-3} \text{ m}^3) = 2.7 \times 10^{25} \text{ m}^{-3}$. We find

$$\tilde{\epsilon} - 1 = 4\pi n_0 \mathcal{L}^3 = 4\pi \cdot 2.7 \times 10^{25} \text{ m}^{-3} \cdot (1.0 \times 10^{-10} \text{ m})^3 = 3.37 \times 10^{-4}. \tag{4.64}$$

The relative electric permittivity of some gases at standard condition are listed below. We write the relative electric permittivity of an ideal gas in terms of an 'effective' radius,

$$(\epsilon_r - 1) = 4\pi n_0 R_{\text{eff}}^3. \tag{4.65}$$

The effective radius in scaled units follows:

$$R_{\text{eff}} = \left(\frac{\epsilon - 1}{\tilde{\epsilon} - 1} \right)^{1/3}. \tag{4.66}$$

In table 4.3 we list the dielectric properties for three noble gases: we see that the effective radius is about twice the radius of the electron cloud, $R_{\text{eff}} \sim 2R$.

4.4.3 Rigidity of atoms

We have focussed on the response of the electrons in atoms due to the force produced by external electric and magnetic fields and found that the electron cloud is displaced by only a small fraction even in a strong external field. The (outer) radius of the electron cloud R determines the size of atoms. Our result thus suggests that the

Table 4.3. Dielectric properties of some gases. R is the radius of the electron cloud, ϵ and R_{eff} is the 'effective' radius calculated from the dielectric constant.

Gas	R	$(\epsilon - 1) \times 10^4$	R_{eff}
He	0.29	0.684	0.59
Ne	0.354	1.27	0.73
Ar	0.659	5.16	1.15

atoms are rigid so that the fractional change radius of the electron cloud is of the order $\delta R/R \sim 10^{-6}$ if an external force is applied. The force can be due to an electric field or have a mechanical origin. This rigidity is the result of the balance between a large attractive and repulsive force that produces a deep and very narrow potential well for electrons. In fact, atoms can be treated as hard spheres (i.e. similar to billiard balls) and is a well-known approximation of atoms in molecular dynamics simulations of liquids [1].

References

[1] Allen M P and Tildesley D J 2017 *Computer Simulations of Liquids* 2nd edn (New York: Oxford University Press)
[2] Bowers M 1986 The classical polarizability of the hydrogen atom *Am. J. Phys.* **54** 347
[3] Goldstein H 1980 *Classical Mechanics* 2nd edn (Reading, MA: Addison Wesley)
[4] Jackson J D 1999 *Classical Electrodynamics* 3rd edn (New York: Wiley)
[5] Landau L D and Lifshitz E M 1976 *Quantum Mechanics: Vol 3 of Course in Theoretical Physics* (Oxford: Butterworth and Heineman)
[6] Pais A 1986 *Inward Bound* (New York: Oxford University Press)
[7] Schwerdtfeger P 2014 Table of experimental and calculated static dipole polarizabilities for the electronic ground states of the neutral elements (in atomic units) http://ctcp.massey.ac.nz/tablepol2014.pdf (Accessed: 7/1/2018)
[8] Silbey R J, Alberty R A and Bawendi M G 2005 *Physical Chemistry* 4th edn (New York: Wiley)
[9] Tabor D 1991 *Gases, Liquids and Solids and Other States of Matter* 3rd edn (New York: Cambridge University Press)
[10] Walecka J D 2008 *Introduction to Modern Physics* (Singapore: World Scientific)

Chapter 5

Properties of small molecules

5.1 Charge distribution in molecules

Neutral atoms have the same number of protons in their nuclei as the number of electrons in 'orbitals', or 'electron shells' that determine atomic radii. The center of shells coincides with the nucleus so that the neutral atoms do not produce an electric field outside the orbitals. However, except for noble gases (group VIII: He, Ne, Ar, Kr, Xe, and Rn (radioactive)) that exist as *monatomic* gases, most chemical elements form *molecules*, that is structures of atoms with fixed distances and angles between them. That is, molecules are described by (chemical) bonds which necessitates the balance of attractive and repulsive forces between atoms.

Noble gases have closed electronic shells so that the atoms remain neutral, and are modeled by a monatomic gas; at room temperatures $T \simeq 300$ K or higher, its properties are described by the ideal gas law $PV = RT$ (for $n = 1$ mol). At lower temperatures, deviations from the ideal gas law behavior become apparent and the ideal gas law is replaced by the van der Waals law $(P + a/V^2)(V - b) = RT$. For argon, the numerical values are $a = 0.1355$ J m^3 mol^{-2} and $b = 3.201 \times 10^{-5}$ m^3 mol^{-1}. The parameters a and b define the energy $RT_c = (8a/27b) = 1.25$ kJ mol^{-1} so that 'critical temperature' is $T_c \simeq 151$ K. For temperatures $T < T_c$, the isotherms define a condensation region in which the gas becomes a liquid in which the atoms are held together by attractive forces. In the case of argon, these attractive forces define a binding energy of the order of the energy scale $E_b \simeq \mathcal{E} = 1$ kJ mol^{-1}.

This binding energy is exceptionally small and, in general, binding energies in molecules are much higher so that (small) molecules can become quite rigid and hard to break up. In fact, the carbon monoxide (CO) molecule is found in the photosphere of the Sun[1]. Since the temperature of the photosphere is about 5000 K, we conclude that the binding energy of the C–O bond is about two order of magnitudes higher than that of the argon gas, $E_b \simeq 100$ in scaled units. There is no interaction between

---

[1] See https://www.britannica.com/place/Sun (retrieved 6/8/2018).

neutral atoms that would yield such a strong bond, and we conclude that the binding of atoms in molecules will generally require some reorganization of charge among atoms. Some atoms become positively charged (cation) while others become negatively charged (anion): the Coulomb force between positive and negative charges provides the attractive force that holds a molecule together. The molecule does not collapse and the distance between ions is fixed (if we ignore vibrations); thus the attractive Coulomb force must be balanced by a repulsive force that reflects the quantum-mechanical nature of electrons (uncertainty and Pauli principle).

The rearrangement of electrons among atoms in molecules is the subject of quantum chemistry, and a detailed description is outside the scope of this book. Here, we discuss qualitatively two simple cases that lead to ionic and covalent bonding, respectively. The nature of most chemical bonds, e.g. the bond between the oxygen and a hydrogen atom in a water molecule, lies 'between' these two extreme cases.

Ionic bonding applies to alkali halides, that is the bond between an alkali atom in group I of the period table (Li, Na, K, ...) and a halogen atom in group VII (F, Cl, Br, ...) when they form a salt (such as *table salt,* NaCl). The alkali atom has one electron outside a closed shell, and the halogen atom misses one electron to form a closed shell. Because closed shells are energetically favorable, one electron will 'move' from the alkali to the halogen atom; as a result the neutral Li and F atoms become positively and negatively charged ions, Li^+ and F^-, respectively. This is shown schematically in figure 5.1. The attractive force responsible for bonding is due to the Coulomb force between the opposite charges on the atoms, $q_{Li} = +1$ and $q_F = -1$. Because the separation between ions is of the order of the characteristic length scale $\mathcal{L}$, so that $a \simeq 1$ in scaled units, the potential energy between the ions follows, $E_C = 1390 q_{Li} q_F / a \simeq -1000$, and thus has the desired order of magnitude.

Covalent bonding describes the situation when two electrons are 'shared' by two atoms, shown schematically in figure 5.2. This applies, for example, to the formation of the diatomic hydrogen molecule H_2. Because the two electrons are in close proximity, the electrons must be in a singlet state: one electron has spin 'up' and the other electron has spin 'down'. If the two ions H^+ (that is, a proton) share both electrons, they each form a closed 1s shell and the electron energy has a minimum.

While covalent bonding requires a quantum-mechanical treatment, we now discuss the electrostatic forces within the H_2 molecule, see figure 5.3. To this end, we assume that the protons are separated by the 'bond length' a; the two electrons are located between two protons so that the distance between protons and the

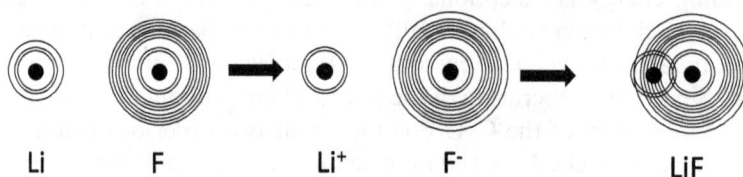

Li   F Li^+ F^- LiF

Figure 5.1. Simplified formation of an alkalihalide. Step 1: Neutral Li and F atoms. Step 2: An electron is transferred from the Li- to the F-atom. Step 3: The cation the cation Li^+ and anion F^- form a diatom.

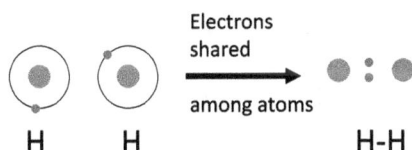

Figure 5.2. Formation of a H_2 molecule from two neutral H atoms.

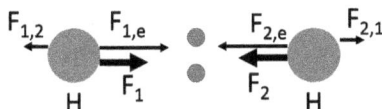

Figure 5.3. Forces in covalent bonding.

electrons is $a/2$. The atoms on the left and right are labeled as '1' and '2' respectively. Because we are interested in the *attractive* forces between the two positively charged protons, we only discuss the forces on the protons, and ignore the forces on the electrons. We use a coordinate system such that 'to the right' is positive (along +x-axis) and 'to the left' is negative (along −x-axis). The force on proton #1 due to the electrons is $F_{1,\,e} = 1390\,(1)\,(2)/(a/2)^2 = 8 \cdot 1390/a^2$; similarly the force on proton #2 due to the two electrons is $F_{2,\,e} = -8 \cdot 1390/a^2$. The two protons repel each other so that $F_{1,2} = -F_{2,1} = -1390 \cdot (1)\,(1)/a^2 = -1390/a^2$. Thus, the net force on proton #1 follows, $F_1 = F_{1,\,e} + F_{1,2} = 7 \cdot 1390/a^2$, and likewise $F_2 = F_{2,\,e} + F_{2,1} = -7 \cdot 1390/a^2$. While the forces $\overrightarrow{F_1}$ and $\overrightarrow{F_2}$ on the two protons are equal in magnitude and opposite in direction, $\overrightarrow{F_1} = -\overrightarrow{F_2}$, they are *not* action–reaction pairs; however they describe an 'effective' mutual interaction between two protons associated with covalent bonding. The corresponding interaction energy is $E_C = -7 \cdot 1390/a \simeq -10\,000/a$.

We conclude that even a small imbalance between positive and negative charges among atoms in a molecule produce Coulomb forces that explain the attractive forces that hold molecules together. This is not sufficient, however, because it would cause the molecule to collapse rather than to attain some finite size with bonds between atoms having a characteristic length. It is necessary, of course, that the attractive electrostatic force is balanced by a repulsive force.

In the case of ionic bonding, we see that the alkali ion Li^+ has two electrons in the 1s-shell that form a 'cloud' which overlaps with the electron cloud of the halogen ion F^-. Because electrons with the same spin (i.e. a triplet state) 'share' the same space, there is an enormous repulsive interaction due to the Pauli exclusion principle. In the case of covalent bonding, the electrons are effectively confined to a small region between the two atoms; that is, the electrons can be described, at least in a rough approximation, by a particle in a box model discussed in section 2.5. The electrons have large positive kinetic energies, which mimic a repulsive force between the atoms.

Attractive forces between atoms, the basis of chemical bonding, are based on the redistribution of electrons from one atom to another and the tendency of atoms.

Pauling introduced the concept of *electronegativity* to quantify the tendency of atoms to attract electrons. Elements with higher electronegativity have higher tendency to attract electrons. The electronegativity for the elements of the first four series of the periodic table are given in table 5.1; we note that the values have uncertainty (based on the model used). We see that for a fixed period, the values increase from group I to group VII. For example, the values for Li and F are 0.98 and 3.98, respectively; this agrees with the formation in alkali halides where the alkali and halogen atoms lose and gain one electron, respectively. The electronegativity of the periodic table of the first four periods of the periodic table is given below; we note that the values vary depending of the model used in the computational method.

The electronegativity of carbon and oxygen are 2.55 and 3.44, respectively, so that the electron 'prefers' the oxygen over the carbon. In general, only a fraction of an elementary charge is transferred from one atom to the other, and the charge on an atom has a fractional value: *partial charge*. For a carbon monoxide molecule, the partial charges are $\delta q_C = + 0.02$ and $\delta q_O = - 0.02$, whereas for the carbon dioxide molecule, the partial charges are $\delta q_C = +1.46$ and for the two oxygen atoms have charge $\delta q_O = - 0.73$. This example shows that partial charges do not characterize individual atoms but rather characterize atoms in a particular molecule. The electronegativity of hydrogen is 2.20, which is lower than the electronegativity of oxygen (3.44) so that in a water molecule, electrons move from the two hydrogens to the oxygen: the partial charges are $\delta q_H = +0.4$ and $\delta q_O = - 0.8$. The partial charge must be computed using a quantum-mechanical description so that the values depend on the model used for the calculation.

It is instructive to discuss chemical bonding in the 'language' of (introductory) chemistry. The valence electrons of an atom are differentiated between 'binding' and 'non-binding' electrons. Two binding electrons form a single bond '–'. Four and six electrons form double and triple bonds, respectively. In the so-called Lewis structure, the bonds and the non-binding electrons are indicated. For example, the Lewis structure of the hydrogen molecule H_2 is written H–H. The neutral carbon and oxygen atoms have four and six *valence* electrons so that the CO molecule has a

Table 5.1. Electronegativity of elements.

Group I	Group II	Group III	Group IV	Group V	Group VI	Group VII
H						
2.20						
Li	Be	B	C	N	O	F
0.98	1.57	2.04	2.55	3.04	3.44	3.98
Na	Mg	Al	Si	P	S	Cl
0.93	1.31	1.61	1.90	2.19	2.58	3.16
K	Ca	Ga	Ge	As	Se	Br
0.82	1.00	1.81	2.01	2.18	2.55	2.96

Carbon monoxide   Carbon dioxide   water

$$:C \equiv O: \qquad .\ddot{O}=C=\ddot{O}. \qquad \overset{..}{\underset{H \quad\quad H}{O}}$$

(a)   (b)   (c)

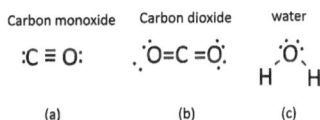

Figure 5.4. The Lewis structure of (a) carbon monoxide CO, (b) carbon dioxide CO_2, and (c) water H_2O.

total of $4 + 6 = 10$ 'shared' electrons. Elementary chemistry shows that there is a triple bond with each bond representing two valence electrons. The Lewis structure is shown in figure 5.4(a). The carbon and oxygen have two unpaired electrons. Thus, the carbon atom has gained one electron in the process of the formation of the molecule, while the oxygen has lost one electron: we say that the carbon has a 'formal charge'[2] of $FC = 4 - \frac{1}{2} \cdot 6 - 2 = -1$ and the oxygen has a formal charge $FC = 6 - \frac{1}{2} \cdot 6 - 2 = +1$. The CO_2 molecule is linear and has $4 + 2 \cdot 6 = 16$ valence electrons. The Lewis structure is shown in figure 5.4(b). The carbon and oxygen atoms have a double bond: there are four unpaired electrons on each of the carbon atoms. The formal charges of carbon of the carbon is $FC = 4 - \frac{1}{2} \cdot 8 - 0 = 0$ and for the two oxygens $FC = 6 - \frac{1}{2} \cdot 4 - 4 = 0$. Two hydrogen atoms and one oxygen atom form a water molecule H_2O. The Lewis structure is shown below in figure 5.4(c). There are two single bonds and the oxygen has four unpaired electrons. Thus the formal charge of the oxygen is $FC = 6 - 2 \cdot \frac{1}{2} \cdot 2 - 4 = 0$ and the formal charge of the two hydrogen atoms is $FC = 1 - \frac{1}{2} \cdot 2 - 0 = 0$. This shows that the term 'charge' has very different meanings; in particular, 'formal' charge is different from 'partial' charge.

5.2 Intramolecular forces

The discussion of the previous section suggests that the main contributions to (chemical) bonding is due to attractive Coulomb forces between positive and negative charges in a molecule and repulsive forces between atoms. We assume that the partial charges for each of the atoms are known (e.g. from a quantum-mechanical calculation) and discuss the electrostatic problem that follows from it. The repulsive force is short ranged and very strong; the exact (analytical) expression is largely irrelevant and is chosen mostly for convenience. In fact, in many numerical calculations, atoms are modeled as hard spheres with some radius a. This corresponds to a (non-analytic) potential with $V_R(r) = \infty$ for $r \leqslant a$ and $V_R(r) = 0$ for $r > a$. We now discuss the various terms that describe attractive and repulsive interactions associated with the bonding of pairs of atoms. We do not discuss bonding that involves more than two atoms; this excludes in particular the 'resonance bonds' in cyclic molecules such as hydrocarbons with an aromatic ring, such as C_6H_6.

---

[2] The formal charge (FC) can be calculated from the equation $FC = N$(valence electrons)$-\frac{1}{2}N$(binding electrons) $- N$(binding electrons).

5.2.1 Attractive forces

Partial charge

We have two atoms with partial charge δq_1 and δq_2 separated by the distance r. The two partial charges have opposite signs, e.g. $\delta q_1 > 0$ and $\delta q_2 < 0$, so that the attractive potential can be written

$$V_C = 1390 \frac{\delta q_1 \cdot \delta q_2}{r}, \tag{5.1}$$

cf. equation (3.58). The Coulomb potential between pair of charges of opposite sign is negative, $V_C < 0$.

Induced dipole moment

We consider a pair of atoms with partial charges $\delta q_1 > 0$ $\delta q_2 < 0$ that are separated by the distance r. Our discussion of the induced dipole moments and their interactions are based on figure 5.5. We choose a coordinate system such that the unit vector $\hat{e}_x$ is aligned with the unit vector $\hat{e}_r$ from atom #1 to atom #2. The charge δq_1 produces an electric field at the location of the second ion δq_2,

$$\overrightarrow{E_1} = \frac{1390 \, \delta q_1}{r^2} \, \hat{e}_x. \tag{5.2}$$

As a result, a dipole moment is induced on atom #2:

$$\overrightarrow{p_2} = \alpha_2 \overrightarrow{E_1} = \alpha_2 \frac{1390 \cdot \delta q_1}{r^2} \, \hat{e}_x. \tag{5.3}$$

Thus, the dipole moment of one atom induced by charge of the other atom decreases with the inverse second law of the distance between the two ions, $p_2 \sim r^{-2}$. We write the polarizability in atomic units (au), cf. equation (4.58), and find $p_2 = [\alpha_2 \cdot (0.529)^3 / 1390] \cdot [1390 \, \delta q_1 / r^2] = \alpha_2 \cdot (0.529)^3 / r^2$ so that

$$\overrightarrow{p_2} = \frac{0.148 \cdot \alpha_2 \delta q_1}{r^2} \cdot (\text{au})^{-1} \hat{e}_x. \tag{5.4}$$

The dipole moment $\overrightarrow{p_2}$ 'feels' the electric field $\overrightarrow{E_1}$ so that the potential energy follows, $V = -\overrightarrow{p_2} \cdot \overrightarrow{E_1} = -[0.148 \, \alpha_2 \delta q_1 / r^2 \, (\text{au})^{-1}] \cdot [1390 \, \delta q_1 / r^2]$, or

$$V = -206 \frac{\alpha_2 (\delta q_1)^2}{r^4} \, (\text{au})^{-1}. \tag{5.5}$$

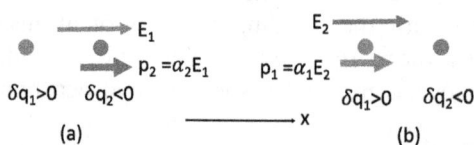

Figure 5.5. Atomic polarization between two atoms with charges $\delta q_1 > 0$ and $\delta q_2 < 0$. (a) Electric field $\overrightarrow{E_1}$ produced by the charge δq_1 and the polarization of atom #2 $\overrightarrow{p_2} = \alpha_2 \overrightarrow{E_1}$. (b) Electric field $\overrightarrow{E_2}$ produced by the charge δq_2 and the polarization of atom #1 $\overrightarrow{p_1} = \alpha_1 \overrightarrow{E_2}$.

We note that the potential energy is negative (the induced force is attractive) independent of the sign of the partial charge δq_1. Likewise the partial charge δq_2 of atom #2 produces an electric field $\overrightarrow{E_2} = -1390 \, \delta q_2/r^2 \cdot \hat{e}_x$ at the position of atom #1. The polarization of the atom #1 is $\overrightarrow{p_1} = \alpha_1 \overrightarrow{E_2} = -0.148\alpha_1\delta q_2/r^4 \cdot (\text{au})^{-1} \hat{e}_x$ and the potential energy is $V = -\overrightarrow{p_1} \cdot \overrightarrow{E_2} = -[0.148 \, \alpha_1\delta q_2/r^2 \, (\text{au})^{-1}] \cdot [1390 \, \delta q_2/r^2]$. We add these two terms and arrive at the expression for the interaction energy due to the induced dipole moments,

$$V_{\text{ind}-\text{dipole}} = -206 \, \frac{[\alpha_1(\delta q_2)^2 + \alpha_2(\delta q_1)^2]}{r^4} \, (\text{au})^{-1}. \tag{5.6}$$

It follows that the potential energy due to the induced dipole moment decreases with the inverse fourth power of the separation between the two atoms, $V_{\text{ind}-\text{dipole}} \sim 1/r^4$.

Dipole–dipole moment
The two induced dipoles $\overrightarrow{p_1}$ and $\overrightarrow{p_2}$ interact with each other via a dipole–dipole interaction term, cf. equation (3.55). Since the two dipole moments are aligned with the line connecting the two atoms $\overrightarrow{p_1}$, $\overrightarrow{p_2} \| \hat{e}_r = \hat{e}_x$, $V_{\text{dipole}-\text{dipole}} = 1390 \cdot 2p_1p_2/r^3$. We insert the expressions for induced dipole moments p_1 and p_2 and find

$$V_{\text{dipole}-\text{dipole}} = 60.8\frac{\alpha_1\alpha_2\delta q_1 \, \delta q_2}{r^7} \, (\text{au})^{-2}. \tag{5.7}$$

Since the partial charges have opposite signs, the dipole–dipole interaction is attractive and $V_{\text{dipole}-\text{dipole}} < 0$.

Electrostatic attraction
The attractive force between atoms in molecules has an electrostatic origin and the corresponding potential is the sum of the contributions from partial charges, induced dipole, and the dipole–dipole: $V_A = V_C + V_{\text{ind}-\text{dipole}} + V_{\text{dipole}-\text{dipole}}$, or

$$V_A = 1390\frac{\delta q_1 \cdot \delta q_2}{r} - 206 \, \frac{[\alpha_1(\delta q_2)^2 + \alpha_2(\delta q_1)^2]}{r^4} \, (\text{au})^{-1} + 60.8\frac{\alpha_1\alpha_2\delta q_1 \, \delta q_2}{r^7} \, (\text{au})^{-2}. \tag{5.8}$$

This expression for the attractive potential is an expansion in inverse powers of the distance between the atoms $1/r^n$ with exponents $n = 1$, $n = 4$, and $n = 7$, respectively.

The expression for V_A is analogous to the multipole expansion of the electrostatic potential of a charge distribution, which requires that the successive terms are *decreasing* in magnitude, $|V_C| > |V_{\text{ind}-\text{dipole}}| > |V_{\text{dipole}-\text{dipole}}|$. In the multipole expansion, this condition is fulfilled since the charges are assumed to be inside a sphere of radius a and the potential is evaluated at a point outside that sphere so that $r > a$ and the expansion parameter is small, $(a/r) < 1$. There is no analogue of a characteristic radius for the expansion in equation (3.41) so that ordering of the magnitudes of the three terms is not unique and depends on the values of the partial charges δq_1 and

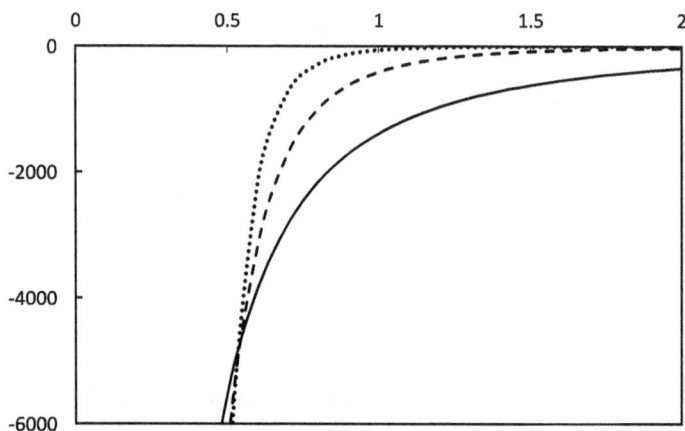

Figure 5.6. The radial dependence of the electrostatic potential due to partial charges (solid), induced dipole moment (dashed), and dipole–dipole interaction (dotted) for $|\delta q_1| = |\delta q_2| = 1$ and $\alpha_1 = \alpha_2 = 1$ au.

δq_2 and the atomic polarizabilities α_1 and α_2. We get a 'feel' for the importance of the various terms by setting $|\delta q_1| = |\delta q_2| = 1$ and $\alpha_1 = \alpha_2 = 1$ au. The radial dependence of the three terms is shown in figure 5.6. We see that for this particular choice of parameters, the three terms have desired order for $r > 0.6$, which is roughly equal to the Bohr radius a_0, i.e. the radius of a hydrogen atom. We caution the reader not to assume that this particular ordering is typical; below we examine these three terms in the case of sodium chloride (NaCl) and find that the dipole–dipole interaction is dominant.

5.2.2 Dispersion forces

So far we have focussed on atoms involved in chemical bonds in molecules and assumed that there are non-zero partial charges on atoms. As mentioned in section 5.1, even argon gas shows deviations from the ideal gas law behavior at sufficiently low temperatures. This suggests that there is an interaction even between neutral atoms with zero partial charges. Neutral atoms are spherically symmetric and the centers of positive and negative charges are identical. However, electrons orbit the atomic nucleus so that the nucleus and the electron cloud form a time-dependent dipole $p_{\text{atom}}(t)$. The time-averaged dipole moment vanishes, $\langle p(t) \rangle = 0$. When two neutral atoms are in proximity to each other, the two atomic dipole moments interact with each other via a dipole–dipole interaction, $V_{\text{atom–atom}} \sim p_1 p_2$. If the two dipole moments are independent of each other, the time-average of the potential energy vanishes, $\langle V_{\text{atom–atom}} \rangle \sim \langle p_1 \rangle \langle p_2 \rangle = 0$.

Because electrons are described by the Schrödinger equation rather than Newton's laws, classical analogs must be interpreted with caution. With this in mind, perhaps the closest analog for the 'motion' of electrons is a random walk (Brownian motion). Thus, the instantaneous dipole moment of an atom is likewise a fluctuating quantity with zero average. Since it is energetically favorable that the dipole moments are aligned with each other, subtle correlations between the two fluctuating dipoles exist that produce an attractive force between the neutral atoms [7]. In the context of the bonding of atoms and molecules, this force is usually referred to as van der Waals force and decays with the sixth power of the separation between the two atoms:

$$V_{\text{vdW}} = -\frac{C}{r^6},$$

where the exponent has to be determined by comparison with experimental or numerical data. The Van der Waals force and the Casimir force between two conducting plates are examples of fluctuation-induced forces [4].

5.2.3 Repulsion

Born and Mayer in 1932 were first to consider the stability of alkalihalide molecules [1] and introduced a short-range repulsive interaction between atoms that captures the Pauli repulsion of electrons in a triplet state (i.e. both spins are up or down). The potential must be stronger than any attractive force at short distance. If the attractive forces have power-law behavior, $V(r) \sim 1/r^n$, for some exponent $n > 0$, a simple functional dependence follows exponential behavior,

$$V_R(r) = V_{R,0} e^{-\kappa r}, \tag{5.9}$$

provided that the parameter κ is sufficiently large. The quantity $1/\kappa$ has the physical dimension of a length and can be identified with the radius (size) of the atom in hard-core approximation $\kappa \sim 1/a$.

Alternatively, the repulsive potential can be written in terms of a power law provided that the exponent is sufficiently large. A popular choice, in particular for the repulsion between neutral atoms, is a power-law behavior with exponent $n = 12$,

$$V_R \sim \frac{V_{R,0}}{r^{12}}, \tag{5.10}$$

The reason for this choice is easy to see when it is combined with the Van der Waals attraction between two neutral atoms so that one arrives at an expression with two terms. This is called the Lennard-Jones potential and is usually written in the form [8]

$$V_{\text{LJ}}(r) = 4\epsilon \left[\left(\frac{\sigma}{r} \right)^{12} - \left(\frac{\sigma}{r} \right)^{6} \right]. \tag{5.11}$$

The equilibrium separation is found by setting the force equal to zero $F = -dV_{\text{LJ}}/dr = 0$:

$$\frac{dV_{\text{LJ}}}{dr} = -\frac{4\epsilon}{\sigma} \left[12 \left(\frac{\sigma}{r} \right)^{13} - 6 \left(\frac{\sigma}{r} \right)^{7} \right] = 0, \tag{5.12}$$

so that

$$R = 2^{1/6}\sigma \simeq 1.12\,\sigma. \tag{5.13}$$

Table 5.2. Lennard-Jones parameters for some gases taken from [8].

gas	ϵ	σ
Ar	1.00	3.41
Xe	1.84	4.10
H_2	0.31	2.93
N_2	0.79	3.70
O_2	0.98	3.58
Cl_2	2.13	4.40
CO_2	1.64	4.30
CH_4	1.23	3.82
C_6H_6	2.02	8.60

Thus the parameter σ is roughly equal to the equilibrium distance. The binding energy follows:

$$E_{\mathrm{LJ}} = -V_{\mathrm{LJ}}(R) = \epsilon. \qquad (5.14)$$

The Lennard-Jones parameters for some gases are given in table 5.2. We note that the potential well has depth of the order $\mathcal{O}(\epsilon) \simeq 1$, which is about three orders of magnitude less than the binding energy of electrons in atoms. This implies that the binding of neutral atoms via the Lennard-Jones potential can be broken up by thermal energies of a few hundred degrees kelvin; on the other hand, thermal effects can be ignored for electronic properties in atoms; electrons can only be excited by the absorption (of emission) of a photon, see our discussion in section 4.3.

We use $E = k_B T$ and equation (2.60) to convert the binding energy into temperature and compare it with the melting and boiling temperatures which are listed in table 5.3. We see that the liquid melts when $T_{\mathrm{melt}} \simeq 0.75 \, T_{\mathrm{bond}}$. At the melting temperature, T_{melt}, the bond is stretched $R_{\mathrm{melt}} \simeq 1.58$ so that

$$\frac{R_{\mathrm{melt}}}{R} \simeq 1.36. \qquad (5.15)$$

That is, melting occurs when the Lennard-Jones bonds are stretched about 1/3 of the equilibrium length of the bond. In agreement with the Lindemann criterion [6], we find that melting is associated with stretching of molecular bonds. Figure 5.7 shows that melting temperature, T_{melt}, and the stretched bond length, R_{melt}.

5.3 Chemical bonds in molecules

We now proceed to explore the nature of chemical bonds in molecules. In quantum chemistry, the energies and lengths of chemical bonds are calculated from first principles [5]. This is not possible in the framework of the theory presented in the text because the atomic interactions depend on unknown parameters. The electrostatic interactions between atoms depend on values of partial charges and atomic

Table 5.3. Temperatures associated with the Lennard-Jones potential: T_{bond} relates to the binding energy, and T_{melt} and T_{boil} are the temperatures of the gas for melting and boiling.

Gas	T_{bond}	T_{melt}	T_{boil}	T_{melt}/T_{bond}
Ar	120	83	87	0.73
Xe	221	161	165	0.75
N_2	95	63	77	0.81
O_2	118	54	90	0.76

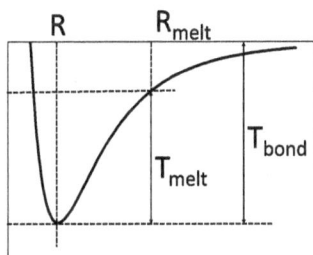

Figure 5.7. The Lennard-Jones potential $V_{LJ}(r)$: R is the equilibrium radius, R_{melt} is the bond stretch when melting occurs at the temperature T_{melt}. The bond energy is given in term of the temperature T_{bond}.

polarizabilities that must be calculated from a full quantum-chemistry treatment. The parameters of the repulsive Born–Mayer potential reflect the quantum-mechanical repulsion of electron clouds and must therefore be obtained by solving a many-electron Schrödinger equation.

We follow an empirical approach explained in the literature [3]. We assume that the electric properties of atoms (i.e. the partial charges δq and the polarizabilities α) are known so that the various terms of the electrostatic interactions between the atoms of a bond can be computed. We then start from empirical values for the binding energies and length of the bond to determine the two parameters characterizing the Born–Mayer potential. While this is not a rigorous treatment, it provides us with insights into the nature of the chemical bond.

5.3.1 Sodium chloride

Sodium chloride is an alkalihalide, and thus one might suspect that the partial charges are $\delta q_{Na} = +1$ and $\delta q_{Cl} = -1$ and the formation of the molecule is described by an ionic bond. In fact, the bonding is not entirely ionic and approximate values for the partial charges are $\delta q_{Na} = +0.8$ and $\delta q_{Cl} = -0.8$. The atomic polarizabilities are found, $\alpha_{Na} \simeq 160$ au and $\alpha_{Cl} \simeq 15$ au. The electrostatic energy for the Na–Cl bond follows:

$$V_A(r) = -\frac{890}{r} - \frac{23\,070}{r^4} - \frac{93\,390}{r^7}. \tag{5.16}$$

The three terms of the attractive potential are shown in figure 5.8.

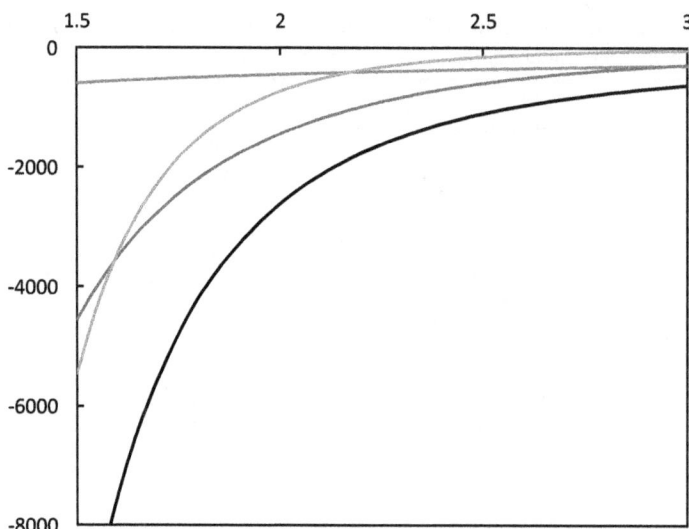

Figure 5.8. Electrostatic potential energy of the alkalihalide NaCl as a function of the Na–Cl distance r. Shown are the Coulomb term $-890/r$ (red), induced dipole term $-23\,070/r^4$ (blue), and dipole–dipole term $-93\,390/r^7$ (green), and the total attractive potential V_A (black).

We assume that the repulsion can be described by a Born–Mayer potential. We thus arrive at the expression for the total energy of the bond:

$$E(r) = V_R(r) + V_A(r) = V_{R,0}e^{-\kappa r} + V_A(r). \tag{5.17}$$

In this expression, the parameters $V_{R,0}$ and κ characterizing the Born–Mayer potential are not known and are determined by comparison with experiments.

Experimental data: For a molecule in the gas phase, the equilibrium length is $r_{eq} = 0.236$ nm and the observed frequency is $f = 388$ cm^{-1}. The dissociation energy of a salt molecule (i.e. the binding energy) NaCl into *neutral* atoms Na and Cl is $E_B = 445$ kJ mol^{-1}. We note that the ionization energy of sodium is 498 kJ mol^{-1} and the molar electron affinity of chlorine is 347 kJ mol^{-1}. We use scaled units and find that the minimum energy of the ion pair Na$^+$ and Cl$^-$ is

$$E_{eq} = -445 + 498 - 347 = -596. \tag{5.18}$$

The molar masses of Na and Cl are $M_{Na} = 22.9$ and $M_{Cl} = 35.5$, respectively. The reduced mass follows:

$$\frac{1}{\mu} = \frac{1}{22.9} + \frac{1}{35.5} = \frac{1}{13.9}. \tag{5.19}$$

In chapter 2, we used the vibrational frequency of sodium chloride as an example for the conversion of frequency in units cm^{-1} to scaled units and found $f = 1.164$ so that the angular frequency follows, $\omega = 2\pi f = 7.31$. The spring constant then follows:

$$k_{NaCl} = \mu\omega^2 = 744. \tag{5.20}$$

Thus the stiffness of the spring that describes the interaction of atoms in chemical bonds is about the same as the stiffness of the spring that describes the interactions of electrons in atoms.

We arrive at this conclusion from spectroscopic data: the UV–vis spectrum includes frequencies from 200 cm^{-1} to 40 cm^{-1}, whereas frequencies in the IR part are lower and range from 40 cm^{-1} to 0.04 cm^{-1}; see table 2.3. UV–vis spectroscopy probes mostly properties of electrons in atoms, whereas IR spectroscopy probes vibrations of atoms in molecules [2].

Since the attractive electrostatic potential energy is known, we write the total energy:

$$E(r) = [V_{R,0} e^{-\kappa r} + V_A(r)], \tag{5.21}$$

where the parameters $V_{R,0}$ and κ for the repulsive Born–Mayer potential are not known and need to be determined from the available data. To this end, we write the equilibrium condition $dE/dr = 0$, or

$$-\kappa V_{R,0} e^{-\kappa r} + V_A'(r) = 0, \tag{5.22}$$

where the prime denotes the derivative with respect to the radius r. The spring constant is determined by the curvature of the energy surface so that

$$\kappa^2 V_{R,0} e^{-\kappa r} + V_A''(r) = 744. \tag{5.23}$$

We note that equations (5.20) and (5.21) are two equations for the two unknowns, $V_{R,0}$ and κ, that can be solved for any value of the radius. We thus find the two parameters for any value of r. We get from equation (5.20) $\kappa V_{R,0} e^{-\kappa r} = V_A'(r)$ so that $\kappa V_A'(r) + V_A''(r) = 744$, or

$$\kappa(r) = \frac{744 - V_A''(r)}{V_A'(r)}. \tag{5.24}$$

Inserted into equation (5.20), we find

$$V_{R,0}(r) = \frac{V_A'(r)}{\kappa(r)} e^{\kappa(r)r}. \tag{5.25}$$

The dependence of the inverse length κ and the strength $V_{R,0}$ as a function of the Na–Cl bond length are shown in figure 5.9. We note that the values of κ and $V_{R,0}$ have a minimum near $r = r_{eq}$.

We proceed by inserting $\kappa(r)$ and $V_{R,0}(r)$ into the expression for the energy, cf. equation (5.19),

$$E_{tot}(r) = [V_{R,0}(r) e^{-\kappa(r)r} + V_A(r)]. \tag{5.26}$$

The function $E(r)$ is shown in figure 5.10. Also shown are the vertical line $r = r_{eq}$ and the horizontal line $E = E_{eq}$. We see that in the case of sodium chloride, all three lines intersect in a single point so that

$$E(r_{eq}) = E_{eq}. \tag{5.27}$$

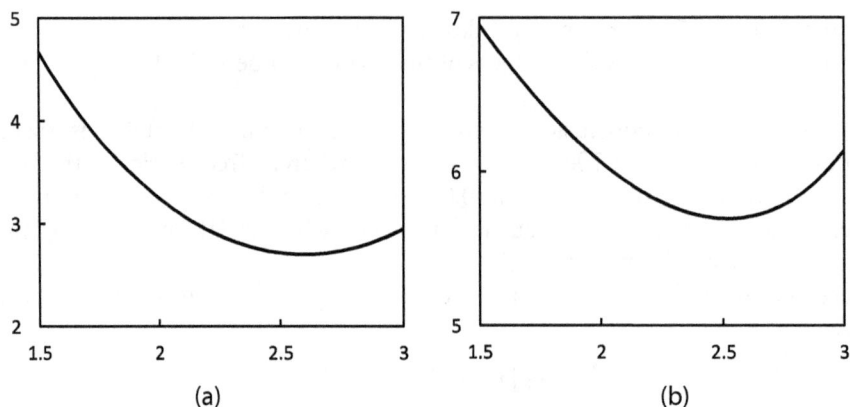

Figure 5.9. The parameters of the repulsive Born–Mayer potential for sodium chloride as a function of the Na–Cl bond length r. (a) The inverse length scale κ, and (b) the strength of the potential log V_{R}, 0.

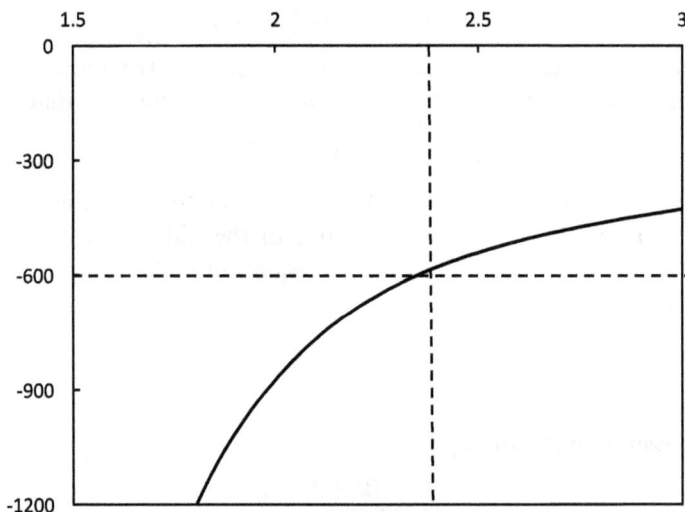

Figure 5.10. The total energy of the NaCl molecule as a function of the bond length R. The dashed vertical line is $r = r_{eq}$ and the dashed horizontal line is $E = E_{eq}$. The vertical and horizontal lines intersect at a point on the line $E = E(r)$ so that $E_{eq} = E(r_{eq})$.

We stress that this is a fortunate coincidence and that in general $E(r_{eq}) \neq E_{eq}$. We obtain the values of the parameters of the Born–Mayer potential:

$$\kappa^{*} \simeq 2.8 \qquad V_{R,0}^{*} \simeq 5.5 \times 10^{5}. \qquad (5.28)$$

The repulsive Born–Mayer potential at the equilibrium distance follows, $V_{R}\,(r_{eq}) = 5.5 \times 10^{5} e^{-(2.8 \cdot 2.3)} \simeq 900$.

The radii of the neutral sodium and chlorine atom are $R_{Na} = 1.71$ and $R_{Cl} = 0.73$, respectively, so that $R_{Na} + R_{Cl} = 2.44$. We thus see that the electronic shells of the two ions are only slightly compressed:

$$\frac{r_{eq}}{R_{Na} + R_{Cl}} = \frac{2.35}{2.44} = 0.96; \tag{5.29}$$

that is, the total compression is about 4%.

We insert the values κ^* and $V_{R,0}$ into the Born–Mayer potential and find the energy:

$$\tilde{E}(r) = V_A(r) + \tilde{V}_{R,0}(r). \tag{5.30}$$

In figure 5.11 we show the attractive electrostatic potential energy V_A and the repulsive Born–Mayer potential $\tilde{V}_R(r)$. We see that for radii close to the equilibrium value $r \simeq r_{eq}$, $V_R \simeq -1000$ and $V_A \simeq 500$ so that the value of the total energy is due a partial cancellation of large positive and negative values. At first glance, the behavior of the total energy for radii in the interval $1.5 < r < 4.0$ might be surprising since it does exhibit a clear minimum at the equilibrium distance r_{eq}. We note, however, that the variation of the attractive and repulsive potential is enormous on this range of the radii and the harmonic approximation is only valid for small deviation from the equilibrium distance,

$$\tilde{E}(r) \simeq E_{harm}(r) = E_{eq} + \frac{1}{2}k_{NaCl}(r - r_{eq})^2, \tag{5.31}$$

where $k_{NaCl} = 744$, cf. equation (5.20).

The melting temperature of sodium chloride crystal is $T_{melt} = 800$ K and corresponds to $\delta E = 6.6$ in scaled units. This is a small change of energy. We use the harmonic approximation to calculate the change in the bond length $\delta E = \frac{1}{2}k(\delta r)^2$ so that

$$\delta r = \sqrt{\frac{2 \cdot 6.6}{744}} = 0.13. \tag{5.32}$$

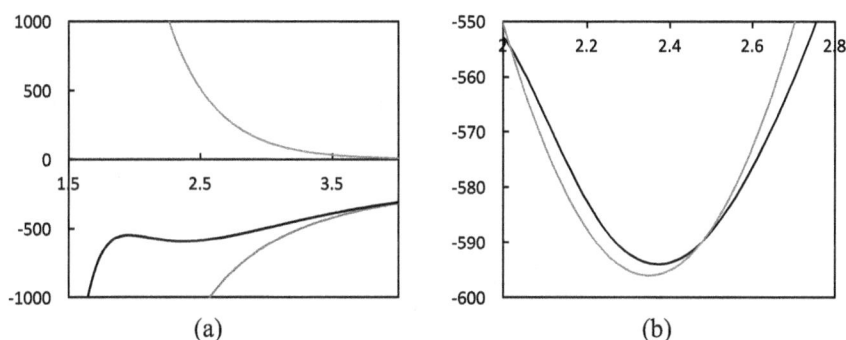

Figure 5.11. The total energy of the NaCl molecule as a function of the bond length R. (a) The attractive electrostatic potential $V_A(r)$ (blue), the repulsive potential $V_R(r)$ (red), and the total energy $E(r) = V_A(r) + V_R(r)$. (b) The total energy $E(r)$ (black) and the harmonic approximation $E_{harm}(r) = E_{min} + \frac{1}{2}k(r - r_{eq})^2$ (red) near the equilibrium distance.

The fractional variation of the bond length follows, $\delta r/r_{eq} = 0.13/2.35 \simeq 6\%$. This estimate is in agreement with the Lindemann criterion which states melting occurs when the (thermal) vibrations of a bond length exceed 10% of the equilibrium length, i.e. $\delta r/r_{eq} \simeq 0.1$.

5.3.2 Water

The water molecule consists of one oxygen atom and two hydrogen molecules. The geometry of the water molecule H_2O is shown in figure 5.12. We now discuss the geometry of the water molecule. We proceed in two steps: (1) we calculate the length of the O–H bond from the attractive electrostatic interaction between the oxygen and hydrogen atom and repulsive potential of the Born–Mayer type; and (2) the angle 2α from the electrostatic repulsion of the two hydrogen atoms and the attractive interaction between two electric dipoles associated with the two O–H bonds.

O–H bond
The partial charges are $\delta q_H = +0.4$ and $\delta q_O = -0.8$. The atomic polarizabilities are $\alpha_H = 4.5$ and $\alpha_O = 6.0$. The electrostatic energy of the O–H bond follows:

$$V_A(r) = -\frac{445}{r} - \frac{791}{r^4} - \frac{525}{r^7}. \tag{5.33}$$

The attractive electrostatic interaction between the oxygen and the hydrogen atoms is shown in figure 5.13.

Experimental data: for water molecule in the gas phase, the equilibrium O–H bond length is $R_{eq} = 0.095$ nm. The 'symmetric' stretch of the water molecule has the frequency $f = 3657$ cm^{-1}. The minimum energy of the O–H bond is found:

$$E_{min} = -460. \tag{5.34}$$

The molar masses of H and O are $M_H = 22.9$ and $M_O = 16$, respectively. The reduced mass follows:

$$\frac{1}{\mu} = \frac{1}{1} + \frac{1}{16} = \frac{1}{0.94}. \tag{5.35}$$

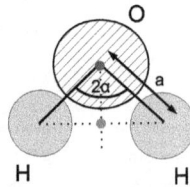

Figure 5.12. The water molecule H_2O; $a = 0.95$ is the bond length and $2\alpha = 109.5°$.

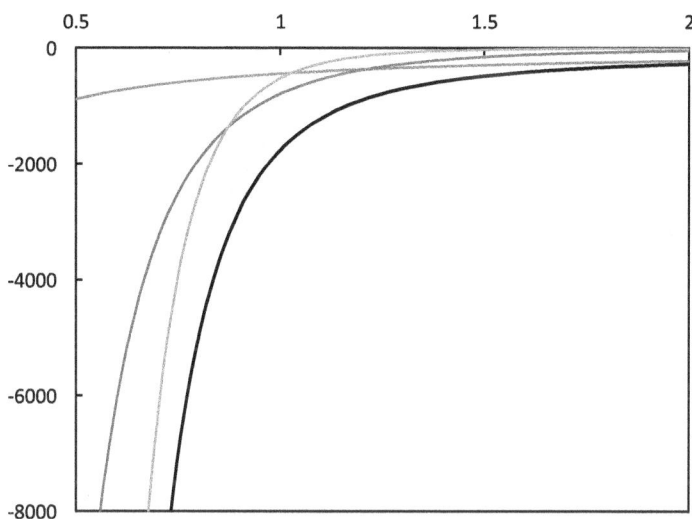

Figure 5.13. Electrostatic potential energy of the O–H bond as a function of the bond length r. Shown are the Coulomb term $-890/r$ (red), induced dipole term $-23\,070/r^4$ (blue), and dipole–dipole term $-93\,390/r^7$ (green), and the total attractive potential V_A (black).

We find the frequency scaled units $f = 10.97$ so that the angular frequency follows, $\omega = 2\pi f = 69.3$. The spring constant of the pseudoforce follows,

$$k_{OH} = \mu\omega^2 = 4460, \tag{5.36}$$

which is comparable to the stiffness of the pseudoforce for the Na–Cl bond.

We proceed as in the case of the NaCl molecule except for changing the changing the electrostatic potential energy and changing the value of the spring constant. The dependence of the inverse length κ and the strength $V_{R,\,0}$ as a function of the O–H bond length are shown in figure 5.14.

The total energy of the O–H bond $E(r)$ as a function of the O–H bond length r is shown in figure 5.15. In this case, the vertical line $r = r_{eq}$, and the horizontal line $E = E_{eq}$ do not intersect at the line $E(r)$ so that $E(r_{eq}) \neq E_{eq}$. At the equilibrium length r_{eq}, we find the energy

$$E(r_{eq}) \simeq -710, \tag{5.37}$$

and the equilibrium energy $E = -480$ corresponds to the radius

$$r(E_{eq}) \simeq 1.3 \tag{5.38}$$

We see from figure 5.14 that the parameters of the Born–Mayer potential have a minimum near $r = 1.15$, which lies in between the experimental value and the estimate obtained from the equilibrium energy $_{eq}$. We thus have

$$\kappa^* \simeq 6.5 \qquad V_{R,\,0}^* \simeq 1 \times 10^6. \tag{5.39}$$

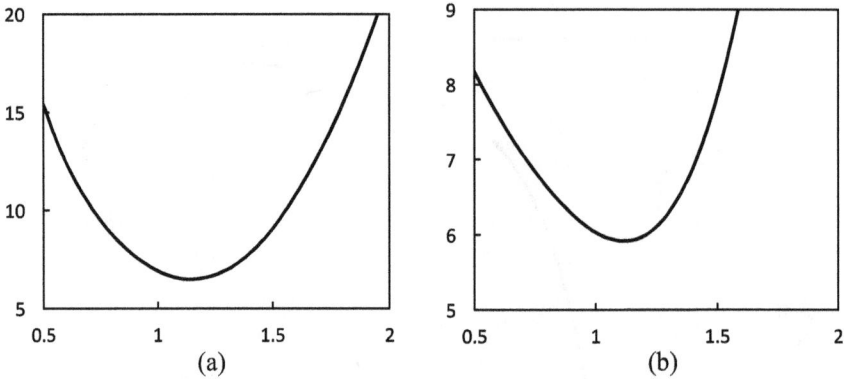

Figure 5.14. The parameters of the repulsive Born–Mayer potential for O–H bond in the water molecule as a function of the O–H bond length. (a) The inverse length scale κ, and (b) the strength of the potential $\log V_{R}$, 0.

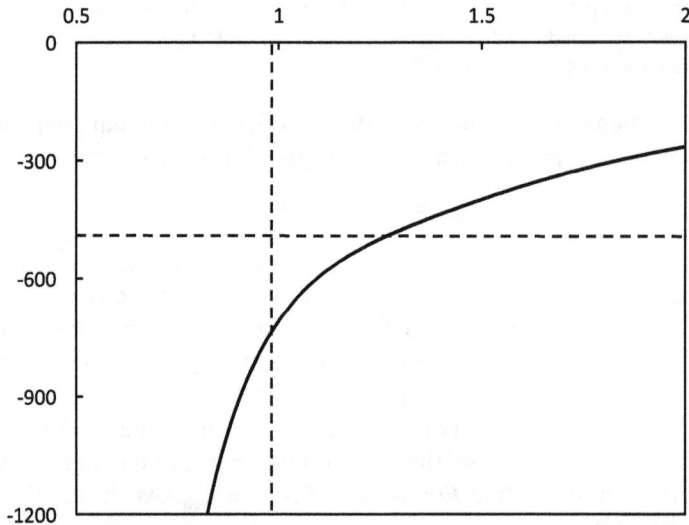

Figure 5.15. The total energy of the O–H bond as a function of the O–H bond length. The dashed vertical line is $r = r_{eq}$ and the dashed horizontal line is $E = E_{eq}$.

The repulsive Born–Mayer potential at the equilibrium distance follows, $V_{R}\,(r_{eq}) = 1.0 \times 10^{6} e^{-(6.5 \cdot 1)} \simeq 1300$.

The radii of the neutral hydrogen and oxygen atoms are $R_{H} = 0.53$ and $R_{O} = 0.45$, respectively, so that $R_{H} + R_{O} = 0.97$. We thus see that the electronic shells of the two ions are not compressed at all:

$$\frac{r_{eq}^{*}}{R_{H} + R_{O}} = \frac{1.35}{0.97} = 1.39. \tag{5.40}$$

The experimental data for bond length O–H is $r_{eq} = 0.95$ so that the ratio follows,

$$\frac{r_{eq}}{R_H + R_O} = \frac{0.95}{0.97} = 0.98; \qquad (5.41)$$

that is, the total compression is about 2%.

We show in figure 5.16 the attractive electrostatic potential energy V_A and the repulsive Born–Mayer potential V_R (r). We see that for radii close to the equilibrium value $r \simeq r_{eq}$, $V_R \simeq -1000$ and $V_A \simeq 500$ so that the value of the total energy is due a partial cancellation of large positive and negative values. The total energy for radii in the interval $0.5 < r < 2.0$ follows the textbook behavior [8]. The harmonic approximation near the equilibrium bond length follows:

$$\tilde{E}(r) \simeq E_{harm}(r) = E_{eq} + \frac{1}{2}k_{OH}(r - r_{eq})^2. \qquad (5.42)$$

We now use the Lindemann criterion to estimate the temperature at which the O–H bonds. Since the equilibrium length is about unity $r_{eq} \simeq 1$, we estimate the thermal vibrations as $\delta r \simeq 0.1$. The corresponding energy follows:

$$\delta E = \frac{1}{2} 4460 \, (0.1)^2 \simeq 22. \qquad (5.43)$$

The temperature (in kelvins) follows:

$$T = \frac{22}{8.3 \times 10^{-3} \text{ K}^{-1}} = 2650 \text{ K}. \qquad (5.44)$$

When the O–H break atomic hydrogen and oxygen are formed, the atoms will combine and for diatomic molecules H_2 and O_2. That is, our estimate shows when the *thermolysis* of water should occur: $2H_2O + \text{heat} \longrightarrow O_2 + 2H_2$. In fact, the value in the literature is $T_{therm} \simeq 2500$ K, and shows that our approximate theory yields surprisingly accurate estimates for the thermal properties of the O–H bond.

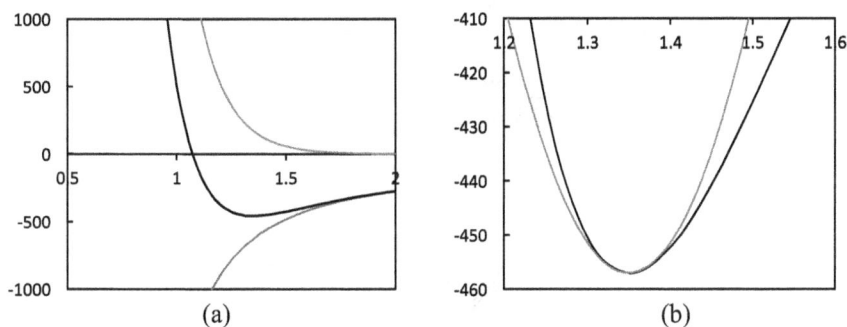

Figure 5.16. (a) The attractive electrostatic potential energy V_A (r) (blue), repulsive Born–Mayer potential (red) and total energy (black) as a function of the O–H distance. (b) The total energy $E(r) = V_A$ $(r) + V_R$ (r) (black) and the harmonic approximation $V_{harm}(r) = E_{eq} + \frac{1}{2}k(r - r_{eq})^2$ (red) as a function of the O–H distance.

O–H bond angles

We assume that the O–H bond length is equal to the equilibrium value $D = 0.95$. We treat the water molecule as two dipole moments $\vec{p_1}$ and $\vec{p_2}$. The magnitude of the dipole moments is

$$|\vec{p_1}| = |\vec{p_2}| = 0.4 \cdot 0.95 = 0.38 = 1.84 \text{ D}, \tag{5.45}$$

where we used the conversion between scaled units and debye, cf. equation (3.59). We now determine the angle between the two O–H bonds, cf. figure 5.17. We show that the angle can be explained by purely electrostatic interactions between the repulsive interactions between the two hydrogen atoms and the attractive interaction between the two dipole moments corresponding to the two bonds.

The two hydrogen atoms carry a positive partial charge; the Coulomb repulsion between the hydrogen atoms tends to increase the angle α. The distance between the hydrogen atoms is $R = 2a \sin \alpha$. The repulsion follows:

$$V_C(\alpha) = 1390\frac{(0.4)^2}{2 \cdot 0.95 \cdot \sin \alpha} = \frac{117}{\sin \alpha}. \tag{5.46}$$

In chapter 3, we derived the expression for the interaction between two 'point dipoles'; this condition is not fulfilled for the dipole model of the water molecule so that our results are only an approximation. We identify the location of the point dipoles with the respective centers of the O–H bonds. For $\alpha = 45°$ the distance between the two dipoles follows, $r \simeq 1/\sqrt{2} \simeq 0.8$. We assume that the 'effective' distance between the dipoles is non-zero even for zero bond angle and then increases monotonously as α increases. We thus arrive at the 'reasonable' estimate for the dipole–dipole distance:

$$r_{\text{dipole–dipole}} \simeq 0.8 + 0.25 \sin \alpha. \tag{5.47}$$

The strength of the dipole–dipole interaction depends on the orientation of the dipoles; we find $\vec{p_1} \cdot \vec{p_2} = p_1 p_2 \cos 2\alpha$ and $\vec{p} \cdot \hat{e}_r = -p_1 \sin \alpha$ and $\vec{p_2} \cdot \vec{e}_r = p_2 \sin \alpha$, cf. figure 5.18. We thus arrive at the expression for the dipole–dipole interaction:

$$V_{\text{dipole–dipole}}(\alpha) \simeq \frac{60.27}{D^2} \frac{[(1.84 \text{ D})^2 \cos 2\alpha - 3(-1.84 \text{ D} \cdot \sin \alpha)(1.84 \text{ D} \cdot \sin \alpha)]}{(0.8 + 0.25 \sin \alpha)^2}$$

$$= 204\frac{[\cos 2\alpha + 3 \sin^2 \alpha]}{(0.8 + 0.25 \sin \alpha)^2}. \tag{5.48}$$

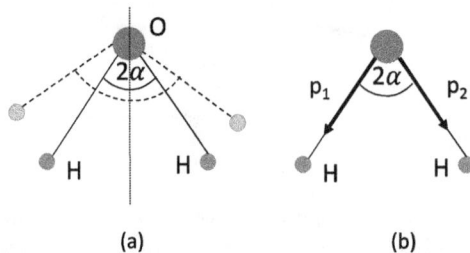

Figure 5.17. The water molecule. (a) The O–H bond lengths are fixed and the angle 2α changes: bending (scissoring) mode. (b) The two O–H bonds correspond to two dipoles that interact with each other.

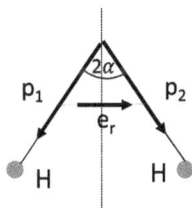

Figure 5.18. Geometry used to find the dipole–dipole interaction that describes the interaction between the two O–H bonds in a water molecule.

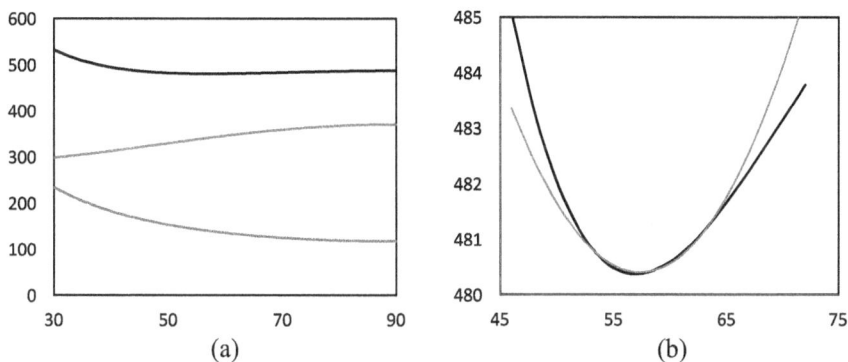

Figure 5.19. (a) The Coulomb repulsion between the two hydrogen atoms (blue), the dipole–dipole interaction (red), and the total energy (black) as a function of the bond angle α. (b) The total energy as a function of the bond angle α and the harmonic approximation (red) as a function of the bond angle α.

Since we assume that the O–H bond lengths are kept fixed, we find the total energy of the water molecule as a function of the bond angle:

$$E(\alpha) = V_C(\alpha) + V_{\text{dipole–dipole}}(\alpha) \simeq \frac{117}{\sin \alpha} + 400\frac{[\cos 2\alpha + 3 \sin^2 \alpha]}{(1 + \sin \alpha/3)^2}. \qquad (5.49)$$

The electrostatic energy as a function of the bond angle α is shown in figure 5.19. We find the equilibrium value of the bond angle:

$$\alpha \simeq 58°, \qquad (5.50)$$

so that $2\alpha_{\text{eq}} \simeq 116$, which is close to the experimental value, $2\alpha_{\text{eq}} \simeq 109.5°$. The harmonic approximation yields the torsion constant κ:

$$E(\alpha) = E_{\text{eq}} + \frac{1}{2}\kappa(\alpha - \alpha_{\text{eq}})^2. \qquad (5.51)$$

We find the value

$$\kappa = 152 \frac{1}{\text{rad}^2}. \qquad (5.52)$$

The moment of inertia of the two hydrogen atoms around the oxygen atom is $I \simeq 2$. Thus, the angular frequency follows,

$$\omega = \sqrt{\frac{\kappa}{I}} = \sqrt{\frac{152}{2}} \simeq 9, \tag{5.53}$$

so that the frequency follows, $f \simeq 1.5$, or $f \simeq 1.5/(3 \times 10^{-3}$ cm$) \simeq 500$ cm^{-1}. The experimental value is $f = 1885$ cm^{-1}, which is about four times our estimate. The poor agreement with experimental results is not surprising, and reflects the sensitivity of the electrostatic interaction on model parameters. In fact, the central result of calculation is the fact that the directional alignment of the water dimer is determined by purely electrostatic interactions.

5.4 Multipole moments

5.4.1 Dipole moment

The experimental value of the O–H bond length in a water molecule is $D = 0.95$ and the angle between the two O–H bonds is $2\alpha = 109°$. The center of the positive partial charges is at the center of the lines connecting the two hydrogen atoms. Thus the center of the positive charges is separated from the negative charge by the distance $D \cos \alpha = 0.95 \cos 54.5° = 0.55$. Since the charges are $2\delta q_H = -\delta q_O = 0.7$, we find the dipole moment

$$p = 0.8 \cdot 0.55 = 0.44 = 2.1 \text{ D}, \tag{5.54}$$

where we used equation (3.59). The actual value is $p = 1.86$ D. The molecule has a zero net charge; that is, the electric *monopole* moment vanishes and the first non-vanishing moment can be the *dipole* moment. This is the case for the diatomic NaCl molecule and the triatomic H_2O-molecule. We say that molecules with a non-vanishing dipole moment are *polar* molecules. Not all molecules are polar, however, and for many molecules, the first non-vanishing multipole moment are the quadrupole, octupole, or any higher-order moment.

5.4.2 Quadrupole moment

We consider the linear CO_2 molecule with partial charges $\delta q_O = -0.73$ and $\delta q_C = +1.46$. The geometry is shown in figure 5.20(a). Each bond has a dipole moment so that the CO_2 molecule consists of two opposite dipole moments, $\vec{p_1} = -\vec{p_2}$, cf. figure 5.20, and the total dipole moment vanishes.

In the linear CO_2 molecule, the electrostatic potential follows:

$$\Phi(r, \theta) = 1390 \frac{\delta q}{r} \left[\frac{2}{r} - \frac{1}{\sqrt{1 - 2(a/r)\cos \theta + (a/r)^2}} \right.$$

$$\left. - \frac{1}{\sqrt{1 + 2(a/r)\cos \theta + (a/r)^2}} \right]. \tag{5.55}$$

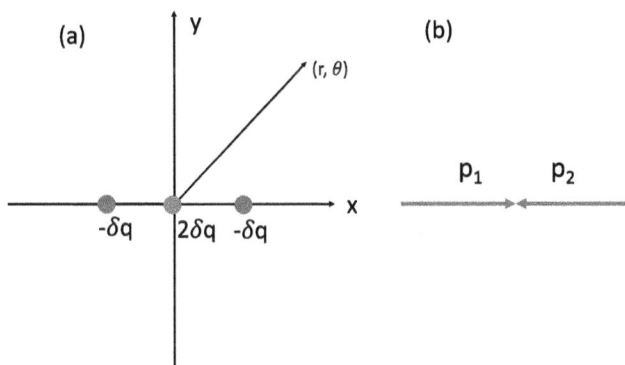

Figure 5.20. The linear CO_2. (a) The molecule lies along the x-axis in the (x, y) plane; the point in the plane has polar coordinates (r, θ). (b) The CO_2 has no dipole moment since $\vec{p} = \vec{p_1} + \vec{p_2} = 0$.

We find

$$\Phi(r, \theta) = 1390 \frac{\delta q}{r} \left[\frac{2}{r} - \sum_{n=0}^{\infty} \{ P_n(\cos \theta) + P_n(-\cos \theta) \} \left(\frac{a}{r} \right)^{n+1} \right]. \qquad (5.56)$$

Since $P_0(\xi) = 1$, the monopole term proportional to (a/r) vanishes. Since $P_{2n+1}(-\xi) = -P_{2n+1}(\xi)$, the odd terms vanish and we get in leading order

$$\Phi(r, \theta) = -1390 \, (3 \cos^2 \theta - 1) \frac{\delta q \cdot a^2}{r^3}, \qquad (5.57)$$

where we used $P_2(\cos \theta) = \frac{1}{2}(3 \cos^2 \theta - 1)$. Comparison with equation (3.41) shows that the electrostatic potential for the CO_2 molecule decays with the inverse third power of the distance $\Phi \sim 1/r^3$. The molecule has zero net charge and zero dipole moment, and the quadrupole term is the first non-zero contribution in the expansion. The quadrupole moment is proportional to the partial charge and the square of the CO distance, $Q_{2m} \sim \delta q \cdot a^2$ (for index $-2 \leqslant m \leqslant 2$).

5.4.3 Octupole moment

We consider the linear CO_2 molecule with partial charges $\delta q_C = -0.35$ and $\delta q_H = +0.09$. The geometry is shown in figure 5.21: the hydrogen atoms occupy the corners of a tetrahedron and the carbon is at the center. We choose the center of the cube as the origin of the coordinate system $(x, y, z) = (0,0,0)$.

The coordinates of the four hydrogen atoms are

$$\vec{r_1} = \begin{pmatrix} a/2 \\ -a/2 \\ -a/2 \end{pmatrix}, \quad \vec{r_2} = \begin{pmatrix} -a/2 \\ a/2 \\ -a/2 \end{pmatrix}, \quad \vec{r_3} = \begin{pmatrix} -a/2 \\ -a/2 \\ a/2 \end{pmatrix}, \quad \vec{r_4} = \begin{pmatrix} a/2 \\ a/2 \\ a/2 \end{pmatrix}. \qquad (5.58)$$

Note that the sum of the x, y, and z components is zero, $\sum x = x_1 + x_2 + x_3 + x_4 = 0$, and likewise for y and z. We calculate the electrostatic potential in the central plane,

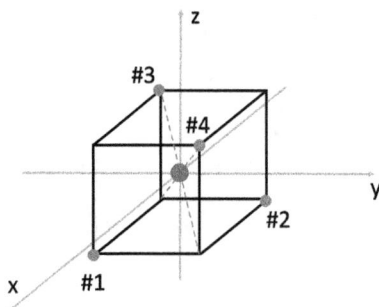

Figure 5.21. The three-dimensional CH_4 molecule. (a) The carbon atoms lie on the corners of a tetrahedron. (b) The leading term of the CH_4 molecule is the octupole term of the multipole expansion.

$z = 0$. We write $x = r \sin\theta \cos\phi$, $y = r \sin\theta \sin\phi$, and $z = r \cos\theta$. Then the electrostatic potential due to the four hydrogen atoms:

$$V_1 = 1390 \, \delta q \frac{1}{\sqrt{(r \sin\theta \cos\phi - a/2)^2 + (r \sin\theta \sin\phi + a/2)^2 + (r \cos\theta + a/2)^2}}$$

$$V_2 = 1390 \, \delta q \frac{1}{\sqrt{(r \sin\theta \cos\phi + a/2)^2 + (r \sin\theta \sin\phi - a/2)^2 + (r \cos\phi + a/2)^2}}$$

$$V_3 = 1390 \, \delta q \frac{1}{\sqrt{(r \sin\theta \cos\phi + a/2)^2 + (r \sin\theta \sin\phi + a/2)^2 + (r \cos\theta - a/2)^2}}$$

$$V_4 = 1390 \, \delta q \frac{1}{\sqrt{(r \sin\theta \cos\phi - a/2)^2 + (r \sin\theta \sin\phi - a/2)^2 + (r \cos\theta - a/2)^2}}$$

and the carbon atom:

$$V_C = 1390 \frac{(-4\delta q)}{r}. \tag{5.59}$$

We introduce $\mathcal{A} = \sqrt{3}/2 \, a$ and $t = \mathcal{A}/r$ and

$$\xi_{\pm, \pm} = \frac{\cos\theta \pm \sin\theta \{\cos\phi \pm \sin\phi\}}{\sqrt{3}} = \frac{\cos\theta \pm \sqrt{2} \sin\theta \cos(\pi/4 \mp \phi)}{\sqrt{3}}. \tag{5.60}$$

We then have for the potential produced by the four hydrogen atoms, $V_H = V_1 + V_2 + V_3 + V_4$:

$$V_H = 1390 \frac{\delta q}{r} \left[\frac{1}{\sqrt{1 + 2\xi_{--}t + t^2}} + \frac{1}{\sqrt{1 + 2\xi_{+-}t + t^2}} \right.$$

$$\left. + \frac{1}{\sqrt{1 - 2\xi_{-+}t + t^2}} + \frac{1}{\sqrt{1 - 2\xi_{++}t + t^2}} \right]$$

The total potential from the methane then follows:

$$V = 1389\frac{\delta q}{r}\left\{\sum_{n=0}^{\infty}[P_n(-\xi_{--}) + P_n(-\xi_{+-}) + P_n(\xi_{-+}) + P_n(\xi_{++})]t^n - 4\right\}. \qquad (5.61)$$

We have $P_0(\xi) = 1$ so that the term $n = 0$ cancels. The term $n = 1$: $P_1(\xi) = \xi$. We find $-\xi_{++} - \xi_{+-} + \xi_{-+} + \xi_{++} = 0$ so that the dipole moment is zero. Likewise, $3[\xi_{--}^2 + \xi_{+-}^2 + \xi_{-+}^2 + \xi_{++}^2] = 4\cos^2\theta + 2\sin^2\theta[2\cos^2\phi + 2\sin^2\phi] = 4$. Since $P_2(\xi) = \frac{1}{2}[3\xi^2 - 1]$, the quadrupole term is zero. For the octupole moment $P_3(\xi) = \frac{1}{2}(5\xi^3 - 3\xi)$. We have

$$[\xi_{--}^3 + \xi_{+-}^3 + \xi_{-+}^3 + \xi_{++}^3] = \cos\theta(5 + 7\sin^2\theta). \qquad (5.62)$$

The electrostatic potential of the methane molecule follows:

$$\Phi(r, \theta, \phi) \simeq 1390\cos\theta(5 + 7\sin^2\theta)\frac{\delta q \, a^3}{r^4}. \qquad (5.63)$$

Comparison with equation (3.41) shows that the electrostatic potential for the CH_4 molecule decays with the inverse fourth power of the distance $\Phi \sim 1/r^4$. The molecule has zero net charge and zero dipole moment, a zero quadrupole term, and the *octupole* term is the first non-zero contribution in the expansion. The octupole moment is proportional to the partial charge and the third power of the CO distance, $Q_{3m} \sim \delta q \cdot a^3$ (for index $-3 \leqslant m \leqslant 3$).

References

[1] Born M and Mayer J E 1932 Lattice theory of ion crystals *Z. Phys.* **75** 1–18 (in German)

[2] Engel T 2006 *Quantum Chemistry & Spectroscopy* 2nd edn (New York: Prentice Hall)

[3] Gowda B T and Benson S W 1982 Empirical potential parameters for alkali halide molecules and crystals, hydrogen alide molecules, alkali metal dimers, and hydrogen and halogen molecules *J. Phys. Chem.* **88** 847–57

[4] Kardar M and Golestanian R 1999 The 'friction' of vacuum, and other fluctuation-induced forces *Rev. Mod. Phys.* **71** 1233–45

[5] Levine I N 2013 *Quantum Chemistry* 7th edn (San Francisco, CA: Pearson)

[6] Lindemann F 1910 The calculation of molecular vibration frequencies *Z. Phys.* **11** 1–18 (in German)

[7] Margenau H 1939 Van der Waals forces *Rev. Mod. Phys.* **11** 1–35

[8] Silbey R J, Alberty R A and Bawendi M G 2005 *Physical Chemistry* 4th edn (New York: Wiley)

Chapter 6

Intermolecular forces

6.1 Formation of crystals

We have seen in the preceding chapter that the interactions between atoms in molecules, i.e. the chemical bonds, are strong so that molecules are very rigid. We have related this property to the fact that electrons redistribute when atoms are in close proximity to each other and form molecules. Atoms acquire (partial) charges and the centers of the electron clouds are shifted away from their corresponding atomic nuclei. As a result, strong attractive forces exist between atoms in a molecule which are balanced by repulsive forces. This rigidity implies that the molecules remain in general electrically neutral unless they a stripped of an electron in an experiment (e.g. in mass spectroscopy). Because the charges are zero, one might suspect that electrostatic forces are no longer important on intermolecular length scales, that is, on length scales of the order of a *nanometer*.

We show in this chapter that this, however, is not the case, since a zero net charge only implies that molecules do not interact via a force with magnitude proportional to the inverse-square power of the distance between molecules, $F_{l=0} \sim 1/r^2$. In general, neutral molecules possess a non-trivial charge distribution associated with the partial charges of the atoms in molecules. The net charge is the monopole term ($l = 0$) of the multipole expansion, cf. equation (3.41), so that the next leading term is the dipole term ($l = 1$). The interactions between two dipoles follows the inverse-fourth power between molecules, $F_{l=1} \sim 1/r^4$.

Both forces are power laws and thus have infinite range and do not define a characteristic length scale. In contrast, the force associated with the Born–Mayer, $F_R(r) = -dV_R/dr = \kappa V_{R,\,0} \exp(-\kappa r)$, defines the characteristic length $\langle r \rangle = 1/\kappa$. We recall that the power-law behavior of the force is only valid outside a sphere of radius r_0 that contains the charge distribution. In general, we define the characteristic length scale as the average distance weighted by the force:

$$\langle r \rangle = \frac{\int_{r_0}^{R} r F(r) dr}{\int_{r_0}^{R} F(r) dr}. \tag{6.1}$$

We set $r_0 = 0$ and recover $\langle r \rangle = 1/\kappa$ for the repulsive force. We assume power-law behavior and write $F(r) = F_R(\rho/r)^{2(l+1)}$ and find

$$\langle r \rangle = \left(1 + \frac{1}{2l}\right) r_0, \qquad l \geqslant 1. \tag{6.2}$$

Here the dependence on the lower radius cutoff r_0 tells us is that the power-law behavior does not have an intrinsic scale. The average distance does not exist for Coulomb's law $l = 0$; indeed, we have

$$\int_{r_0}^{R} r F(r) dr = \int_{r_0}^{R} r F_0 \left(\frac{\rho}{r}\right)^2 dr = F_0 \rho \int_{r_0/\rho}^{R/\rho} \frac{du}{u} = F_0 \rho \ln\left(\frac{R}{r_0}\right) \longrightarrow \infty \tag{6.3}$$

for $R \longrightarrow \infty$. This property of the Coulomb interaction has important implications for macroscopic systems.

We calculate the ratio of the force at r_0 and $\langle r \rangle$ and find $F(r_0)/F(\langle r \rangle) = (1 + 1/2l)^{-2(l+1)}$ so that $F(r_0)/F(\langle r \rangle) \simeq 0.4$ for the dipole term $l = 1$ and $F(r_0)/F(\langle r \rangle) \simeq 1$ for higher dipoles $l \geqslant 2$. If we associate the lower cutoff with the size of a molecule $r_0 = a$ and write the distance as a multiple of a, $r = \nu a$, so that $\nu = 1$ for the nearest neighbor and $\nu = 2$ the next-nearest neighbor and so on. It follows that the multipole interaction between neighbors decays with a power law $F_\nu = F_0/\nu^{2(l+1)}$. The number of neighbors N_ν is proportional to the surface area of a sphere with radius $r = \nu a$ so that we find quadratic behavior, $N_\nu = N_0 \nu^2$. The total force is given by the product of the interaction F_ν and the number of interactions $F_{total} = N_\nu F_\nu \sim \nu^2 \cdot \nu^{-2(l+1)} = 1/\nu^{2l}$. For the dipole–dipole interaction, we thus expect that the total force from all the next-nearest neighbors is only about a quarter $(1/2^2)$ of the interaction from the nearest neighbors. If the interaction is due to the Coulomb force (corresponding to the monopole term $l = 0$), we have $F_\nu \sim \nu^{-2}$ so that the total force follows, $F_{total} = N_\nu F_\nu \sim \nu^2 \cdot \nu^{-2} = const$; that is, the total force from the next-nearest neighbors is the same as the force from the nearest neighbor. Because there are infinitely many relations between molecules, $\nu = 0, 1, 2, \ldots$, we conclude that the total force from an infinite number of point charges would diverge[1].

Our result shows the subsystems of a macroscopic system have zero net charge. This is the case for crystals that are an arrangement of molecules: molecules have zero net charge and intermolecular forces have finite range. One molecule only interacts with its nearest neighbor, next-nearest neighbor, etc., but not with any

---

[1] An analogous divergence is familiar from astronomy. The intensity of stars decays with the inverse-square power of their distance $I \sim 1/r^2$ and the number of stars increases with the square of the distance $N(r) \sim r^2$. If the Universe is assumed to be infinite, this would imply that the night sky is bright. This is known as 'Olber's paradox' or the 'dark night paradox'; see [3].

neighbor that is 'infinitely' far away. This is at first very surprising since molecules can form crystals so that the location and orientation of molecules exhibit long-range correlations. However, these correlations are explained by the laws of statistical physics, in particular the theory of phase transitions [6, 9], and only exist in the thermodynamic limit, i.e. in the case when the number of particles becomes infinitely large. We say that the formation of crystals is an *emergent phenomena* [11]. We recall that Avogadro's number, $N_A \simeq 6 \times 10^{23} \text{mol}^{-1}$, is 'infinitely' large so that even some fraction of one gram of water can form an ice crystal. The transformation of liquid water to solid ice is associated with latent heat, and is therefore called 'first order' in Ehrenfest's classification. The short-range character of intermolecular forces suggests that the molar latent can be explained by the breaking of molecular bonds; in the simplest approximation, we have only to take into account the interactions between nearest neighbors. Theoretical models for phase transitions rarely go beyond nearest-neighbor interactions to also include next-nearest-neighbor interactions; in fact, the celebrated Ising model of phase transitions is also restricted to nearest-neighbor interactions [9].

When a crystal melts, an enormous number of intermolecular bonds must be broken. This phase transformation requires the addition of heat, which is an extensive variable (i.e. proportional to the 'size' of the system). The heat necessary to melt one mole defines the molar latent heat, L_{molar}. The ratio of the molar heat, L_{molar}/N_A, is the energy to remove one molecule from the crystal. Our discussion shows that L_{molar}/N_A is equal to the energy to break the bond of two nearest-neighbor molecules. In the following, we discuss the melting of ice and thus examine the energetics of a water dimer. We note that the properties of small clusters of water is an interesting problem in-itself [5].

6.2 Water dimer

We thus arrive at the much more manageable system of two water molecules. In general, the *intra*-molecular energies are much higher than *inter*-molecular energies; we thus assume that a water molecule is rigid and retains its geometry even in the presence of another molecule. We choose a coordinate system associated with molecule #1: we choose the oxygen atom of molecule #1 as the origin of the coordinate system, and choose the (x, y) so that the two hydrogen atoms of molecule #1 lie in that plane, cf. figure 6.1. The coordinates of the atoms of molecule #1 are given by

$$\vec{r_O} = \begin{pmatrix} 0 \\ 0 \\ 0 \end{pmatrix}, \quad \vec{r_H} = \begin{pmatrix} 0 \\ a \sin \alpha \\ -a \cos \alpha \end{pmatrix}, \quad \vec{r_{H'}} = \begin{pmatrix} 0 \\ -a \sin \alpha \\ -a \cos \alpha \end{pmatrix}, \quad (6.4)$$

where $a = 0.95$ is the equilibrium O–H bond length. We note that the problem is symmetric with respect to the exchange of the hydrogen atoms of molecule #1 so that the problem is symmetric with respect to the reflection about the (x, z) plane.

We know from a general chemistry course that a hydrogen bond is formed when a positively charged hydrogen atom lies between two negatively charged atoms. We expect that the minimum-energy configuration has maximum symmetry: thus, the

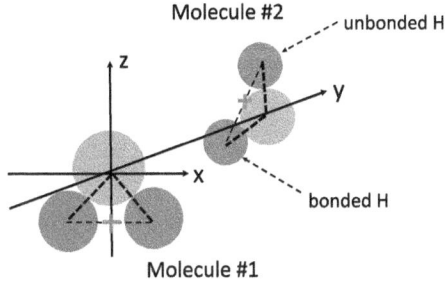

Figure 6.1. Water dimer. The bonded hydrogen lies near the line connecting the two oxygen atoms (taken as the y-axis). Molecule #1 lies in the plane $y = 0$ and molecule #2 lies close to the plane $x = 0$.

oxygen atom of molecule #2 lies along the y-axis at $x = D$, the *bonded* hydrogen atoms of molecule #2 lie at $D - a$, and the *unbonded* hydrogen atom of molecules #2 lies in the (x, z) plane such that the angle between the two O–H bonds is 2α. The orientation of the second molecules is characterized by two angles: the angle γ_1 between the symmetry axis and the x-axis (corresponding to the polar angle for spherical coordinates) and the angle of rotation about the x-axis, γ_2, corresponding to the azimuthal angle for spherical coordinates. We thus arrive at the coordinates for molecule #2:

$$\vec{r}_O = \begin{pmatrix} D \\ 0 \\ 0 \end{pmatrix}, \quad \vec{r}_H = \begin{pmatrix} D - a\cos(\gamma_1 - \alpha) \\ -a\sin(\gamma_1 - \alpha)\sin\gamma_2 \\ a\sin(\gamma_1 - \alpha)\cos\gamma_2 \end{pmatrix}, \quad \vec{r}_{H'} = \begin{pmatrix} D - a\cos(\gamma_1 + \alpha) \\ -a\sin(\gamma_1 + \alpha)\sin\gamma_2 \\ a\sin(\gamma_1 + \alpha)\cos\gamma_2 \end{pmatrix}. \quad (6.5)$$

We assume that molecule #1 is fixed and consider changes in the location and orientation of molecule #2. We first consider changes in the hydrogen bond, i.e. changes in the distance between the two oxygen atoms, and then examine the orientation of molecule #2 for a fixed O–O distance.

6.3 Hydrogen bond

The energy of the hydrogen bond is discussed in detail in reference [12]; we do not repeat the calculation here and summarize them. We use a simplified geometry and use $a \simeq 1$ for the O–H bond length and $2\alpha \simeq 90°$ for the angle between the O–H bonds in the water molecule. Since we are interested in the motion of the oxygen molecule of molecule #2 along the y-axis, we calculate the y-components of the forces on the atoms of molecule #2 due to molecule #1. The force on the oxygen atom is repulsive, $F_{y,O} > 0$, and the forces on the bonded and unbonded hydrogen atoms are attractive, $F_{y,Hb}, F_{y,Hu} < 0$. The magnitude of the forces on the oxygen and unbonded hydrogen atoms are smaller than the force on the bonded hydrogen atom, $|F_{y,O}|, |F_{y,Hu}| < |F_{y,Hb}|$. We find that the sum forces on the oxygen and unbonded hydrogen atoms is much smaller than the force on the bonded hydrogen atom,

$$|F_{y,O} + F_{y,Hu}| \ll |F_{y,Hb}|. \quad (6.6)$$

We find the asymptotic behavior for large separations of the water molecules,

$$F_y(D) \simeq F_{y,\,\text{Hb}}(D) \simeq -\frac{218}{D^5}, \tag{6.7}$$

where the negative sign indicates that the force between the two water molecules is *attractive*. We note the similarity of hydrogen bonding with covalent bonding. In covalent bonding, the attractive force acts between two positively charged atoms, whereas in hydrogen bonding the attractive force is between two negatively charged atoms. The role of the two paired electrons in covalent bonding is replaced by the positively charged bonded hydrogen atom.

We now calculate the electrostatic potential energy of molecule #2 in the electric field produced by the charges on molecule #1. Since $\delta q_O \simeq -0.8 \, \delta q_H + = 0.4$, we find for the oxygen atom,

$$V_O(D) = -1390 \cdot 0.8 \left[\frac{2 \cdot 0.4}{\sqrt{1 + D^2}} - \frac{0.8}{D} \right]. \tag{6.8}$$

We find for the bonded hydrogen,

$$V_{\text{Hb}}(D) = 1390 \cdot 0.4 \left[\frac{2 \cdot 0.4}{\sqrt{1 + (D - 1)^2}} - \frac{0.8}{D - 1} \right], \tag{6.9}$$

and the unbonded hydrogen atom,

$$V_{\text{Hb}}(D) = 1390 \cdot 0.4 \left[\frac{2 \cdot 0.4}{\sqrt{1/2 + D^2 + (1 + 1/\sqrt{2})^2}} - \frac{0.8}{\sqrt{D^2 + 1}} \right]. \tag{6.10}$$

In these expressions, the first term is the contributions from the two hydrogen atoms of molecule #1 and the second term due the oxygen of molecule #1. Overall we find that the electrostatic interaction between the two dimers is *attractive*,

$$V_A(D) = V_O(D) + V_{\text{Hb}}(D) + V_{\text{Hu}}(D) \simeq V_{\text{Hb}}(D). \tag{6.11}$$

The three terms of the electrostatic potential energy are shown in figure 6.2. Since the potential energies from the repulsive interaction between molecule #1 and oxygen atom and the attractive interaction between molecule #1 and the unbonded hydrogen atom nearly cancel each other out, $V_O(D) + V_{\text{Hu}}(D) \simeq 0$, so that the interaction is dominated by the force on the bonded hydrogen atom, $V_A(D) \simeq V_{\text{Hb}}(D)$.

In the geometry of figure 6.1, the two water molecules are attracted to each other. We have seen in chapter 5 that the equilibrium distance of chemical bonds is determined by the radius of the bonded atoms. Thus the O–H distance is roughly equal to the sum of the radii of the oxygen and hydrogen atoms, $R_{\text{OH}} \simeq R_O + R_H$, cf. equation (5.39). We might expect that the two water molecules approach each other until the distance between the oxygen atom of molecule #1 and the bonded

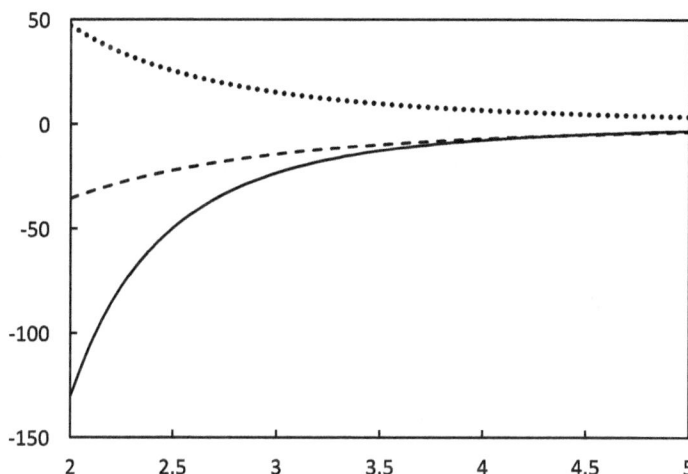

Figure 6.2. The electrostatic potential energies of atoms in molecule #2 due to the electric field produced by molecule #1. The potential energy of the bonded hydrogen atom (solid), the oxygen atom (dashed), and the unbonded hydrogen atom (dotted).

hydrogen atom of molecule #2 is equal to the length of an O–H bond in a water atom, $R_{eq} \simeq 0.95$. If this were the case, the O–O distance of the water dimer would be twice the length of an O–H bond: $D \simeq 2r_{eq} \simeq 2$.

In this case, we would expect that there is a redistribution of electrons among the two water molecules so that the partial charges of the oxygen and hydrogen atoms in water molecule would lose their relevance for the water dimer and, of course, even larger structures (water trimer, etc.). This is, however, not the case: water molecules stay intact as a cluster of water molecules is formed. A quantum-mechanical calculation shows that in the stable configuration of the water dimer, the O–O distance is about $D_{eq} \simeq 3$ [5]. We conclude that the distance between the oxygen atom of molecule #1 and the bonded hydrogen atom of atom #2 is about twice the O–H bond length of a water molecule. Thus the electron cloud of the bonded hydrogen atom is well separated from the electron cloud of the oxygen atom of molecule #1, which effectively prevents any redistribution of charges among molecules.

The equilibrium O–H bond is determined by the balance between electrostatic interaction and quantum-mechanical repulsion of electron 'clouds'. We expect that the equilibrium distance between water dimers is similarly determined by the balance between electrostatic interaction and quantum-mechanical repulsion. The equilibrium O–O distance D_{eq} suggests that the Born–Mayer potential for the water dimer is *softer*, that is, it decays less rapidly with distance.

It is shown [12] that the repulsive potential between is weaker, i.e. the prefactor $V_{R,0}$ is smaller for molecular interactions than for atomic repulsion. We write the repulsive potential for the water dimer, cf. equation (5.9),

$$V_R'(D) = V_{R,0}' e^{-\kappa' D}, \tag{6.12}$$

with

$$\kappa' = \frac{\kappa}{\zeta}, \qquad V'_{R,0} = \frac{V_{R,0}}{\zeta}, \tag{6.13}$$

where $\kappa = 6.5$ and $V_{R,0} = 1 \times 10^6$ are the parameters of the Born–Mayer potential for the O–H bond of a water molecule, cf. equation (5.37).

The energy of the hydrogen bond depends on the O–O distance D and the parameter ζ characterizing the repulsive Born–Mayer potential for the repulsion of the water molecules:

$$E(D; \zeta) = V_A(D) + V_R(D, \zeta). \tag{6.14}$$

We choose a value for ζ so that the interaction energy depends only on the O–O distance D. We find the equilibrium length of the hydrogen bond from

$$\left. \frac{dE}{dD} \right|_{D_{eq}} = 0. \tag{6.15}$$

The bonding energy then follows:

$$E_{eq} = E(D_{eq}). \tag{6.16}$$

Note that D_{eq} and E_{eq} thus implicitly depend on the parameter ζ. We vary the parameter $1.5 \leqslant \zeta \leqslant 2.5$ and the determine the equilibrium; the results are listed in table 6.1 and are shown in figure 6.3. It shows that small variation of the parameters ζ of the repulsive potential can change the values of D_{eq} and E_{eq}.

We choose $\zeta = 2$ as a reasonable value and find the corresponding energy,

$$\tilde{E}(D) = E(D; \zeta = 2) = V_A(D) + V_R(D, \zeta = 2), \tag{6.17}$$

Table 6.1. The equilibrium distance D_{eq} and equilibrium energy E_{eq} of the hydrogen bond for a water dimer.

ζ	D_{eq}	E_{eq}
1.5	2.4	−34.6
1.6	2.6	−23.5
1.7	2.8	−16.4
1.8	3.2	−12.4
1.9	3.5	−9.4
2.0	3.7	−7.4
2.1	3.9	−5.9
2.2	4.2	−4.8
2.3	4.6	−3.9
2.4	4.6	−3.2
2.5	4.7	−2.5

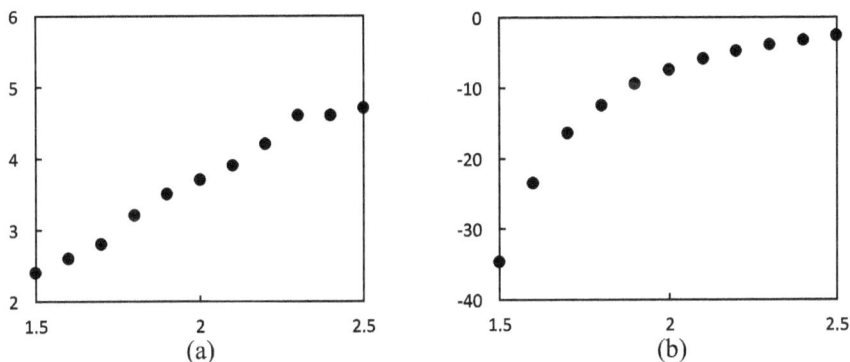

Figure 6.3. The equilibrium values of the hydrogen bond as a function of the parameter ζ. (a) Equilibrium distance r_{eq} and (b) equilibrium energy E_{eq}.

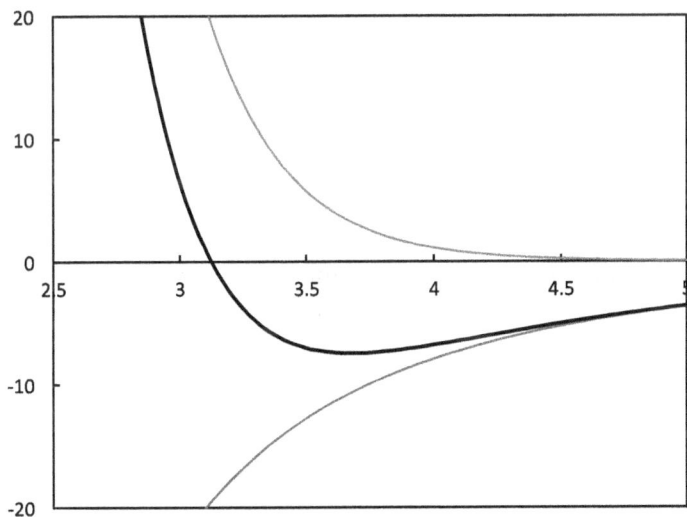

Figure 6.4. The energies of the hydrogen bond of the water dimer as a function of the O–O distance D: electrostatic potential energy V_A (blue), Born–Mayer repulsive potential V_R (red), and total energy $\tilde{E}$ (black).

these terms are shown in figure 6.4. We approximate the hydrogen bond with a spring-like pseudoforce. We use a harmonic approximation of the energy $\tilde{E}(D) \simeq \tilde{E}_{harm}(D)$,

$$\tilde{E}(D) \simeq \tilde{E}_{harm}(D) = E_{eq} + \frac{1}{2}30(D - D_{eq})^2, \tag{6.18}$$

so that the spring constant corresponding to the pseudoforce of the hydrogen bond follows:

$$k_{H-bond} \simeq 30. \tag{6.19}$$

The energy $\tilde{E}(D)$ and the quadratic approximation $E_{harm}(D)$ are shown in figure 6.5.

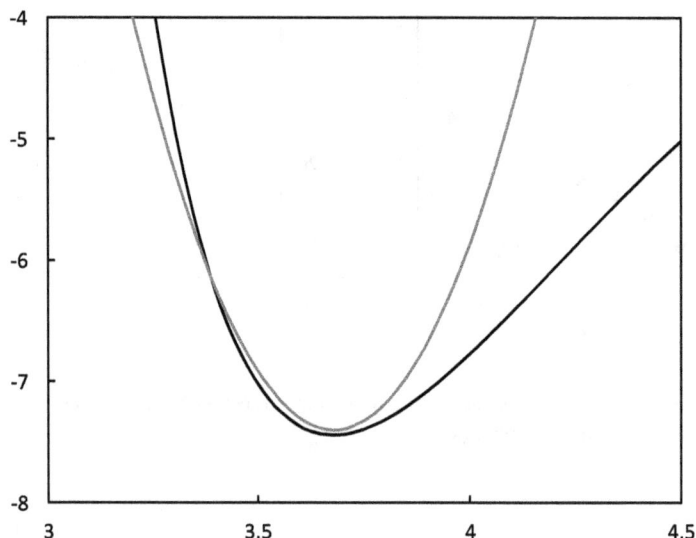

Figure 6.5. The energy of the hydrogen bond of the water dimer as a function of the O–O distance: the total energy $\tilde{E}(D)$ (black) and the harmonic approximation $\tilde{E}_{\mathrm{harm}}(D)$.

We recall from chapter 5 that the spring constants of the pseudoforces corresponding to the chemical bonds in sodium chloride and water molecules are $k_{\mathrm{NaCl}} \simeq 744$ and $k_{\mathrm{OH}} \simeq 4460$, respectively. The interaction between atoms in a molecule corresponds to *intra*-molecular forces, whereas the interaction between water molecules in a dimer corresponds to *inter*-molecular forces. We conclude that springs corresponding to intramolecular pseudoforces are stiffer than those corresponding to intermolecular pseudoforces,

$$k_{\mathrm{intra}} \gg k_{\mathrm{inter}}. \tag{6.20}$$

This result is not surprising and reflects the different nature of the electrostatic interaction between a pair of molecules and a pair of atoms. Atoms are charged and interact with each other via the strong Coulomb force. On the other hand, molecules have no net charge so that attractive and repulsive Coulomb forces partially cancel each other so that a much weaker dipole–dipole (or a higher multipole moment) interaction remains. We note that the Lennard-Jones potential between neutral atoms is also weak, cf. table 5.2. We conclude that the molecules can be treated as *rigid objects* when the assembly into larger structures (dimers, trimers, ..., cluster, and crystals) is examined.

We examine the energies necessary to form structures of molecules. Since the equilibrium O–O distance is $D \simeq 3.5$, Lindemann's criterion [7] states that thermal fluctuations must stretch the hydrogen bond and the (average) O–O distance increases, $\Delta D = 0.35$. We find the change of the energy of the hydrogen,

$$\Delta E \simeq \frac{1}{2} 30 \cdot (0.35)^2 \simeq 2. \tag{6.21}$$

This energy is of the order of thermal energies at room temperatures. We use equation (2.61) to find the temperature of melting:

$$T_{melt} = \frac{2}{8.3 \times 10^{-3}\text{K}^{-1}} \simeq 240 \text{ K}. \tag{6.22}$$

The experimental value is, of course, $T = 273$ K (corresponding to $T_{melt} = 0$ °C). Thus our estimate is correct within about 10% of the empirical value; this is a surprisingly good estimate considering the approximate nature of our theory and the 'guesswork' used to find the parameters of the Born–Mayer potential.

6.4 Dimer orientation

We now turn to the relative orientation of the two water molecules. We use equations (6.4) and (6.5) for the location of the atoms in molecule #1 and #2. The total potential energy of the interaction is given by $V = V_O + V_{Hb} + V_{Hu}$. The generalization of equations (6.9)–(6.11) to case when the two molecules are not perfectly aligned is straightforward, although somewhat tedious. We introduce the parameter $\xi = a/D$ so that $\xi \to 0$ corresponds to the case when the two molecules are far apart and $\xi \to 1$ corresponds to the case when the two molecules are close to each other. Because the reorientation of molecule #2 does not significantly change the distances between atoms, we ignore the variation of the repulsive Born–Mayer potential.

We thus arrive at the expression for the potential energy:

$$
\begin{aligned}
V(\gamma_1, \gamma_2; D, \xi) = \frac{888}{D}\Bigg\{ & 1 - \frac{1/2}{\sqrt{1 - 2\xi\cos(\gamma_1 - \alpha) + \xi^2}} \\
& - \frac{1/2}{\sqrt{1 - 2\xi\cos(\gamma_1 + \alpha) + \xi^2}}\Bigg\} \\
& + \Bigg\{ -\frac{1/2}{\sqrt{1 + \xi^2}} \\
& + \frac{1/4}{\sqrt{1 - 2\xi\cos(\gamma_1 - \alpha) + 2\xi^2[1 + \sin(\gamma_1 - \alpha)\cos(\gamma_2 - \alpha)]}} \\
& + \frac{1/4}{\sqrt{1 - 2\xi\cos(\gamma_1 - \alpha) + 2\xi^2[1 + \sin(\gamma_1 - \alpha)\cos(\gamma_2 + \alpha)]}}\Bigg\} \\
& + \Bigg\{ -\frac{1/2}{\sqrt{1 + \xi^2}} \\
& + \frac{1/4}{\sqrt{1 - 2\cos(\gamma_1 + \alpha) + 2\xi^2[1 + \sin(\gamma_1 + \alpha)\cos(\gamma_2 - \alpha)]}} \\
& + \frac{1/4}{\sqrt{1 - 2\xi\cos(\gamma_1 + \alpha) + 2\xi^2[1 + \sin(\gamma_1 + \alpha)\cos(\gamma_2 + \alpha)]}}\Bigg\}.
\end{aligned}
\tag{6.23}
$$

We focus on the dependence of the intramolecular interaction on the relative orientation of the water molecules. We expect that the angular dependence is more 'pronounced' when the two water molecules are closer to each other, that is, when the distance between two oxygen atoms is similar to the O–H bond length. A direct comparison of the potential at various distances D is obscured by the overall dependence of the potential on the separation of the two molecules, $V \sim D^{-1}$. We remove this trivial dependence by considering the scaled potential,

$$\tilde{V}(\gamma_1, \gamma_2) = \frac{D}{\xi^2} V(\gamma_1, \gamma_2; D, \xi). \tag{6.24}$$

In this fashion, the overall magnitude of $\tilde{V}$ is independent of both D and ξ so that the variation of $\tilde{V}$ reflects the dependence of the interaction on the relative orientation of the water molecules.

(1) We set $\gamma_2 = 0$ so that molecule #2 is confined to the plane $x = 0$. The dynamics $\gamma_1 = \gamma_1(t)$ describes a rocking motion of molecule #2 relative to molecule #1. We write

$$\tilde{V}_1(\gamma_1) = \tilde{V}(\gamma_1, \gamma_2 = 0). \tag{6.25}$$

Figure 6.6 shows the angular dependence of the scaled potential $\tilde{V}_1(\gamma_1)$ for five different separations of the water molecules.

For $\xi \to 0$ and the water molecules are far apart from each other. The potential energy $\tilde{V}_1(\gamma_1; \xi = 0.1)$ has a minimum for $\gamma_1 \simeq 90°$ and a maximum for $\gamma_1 \simeq 270°$. We see that the potential energy has the approximate form of a sine-function $\tilde{V}_1(\gamma_1, \xi = 0.1) \sim -\sin(\gamma_1)$. This shows that the electrostatic interaction between two water molecules can be treated as an interaction between two electric dipoles.

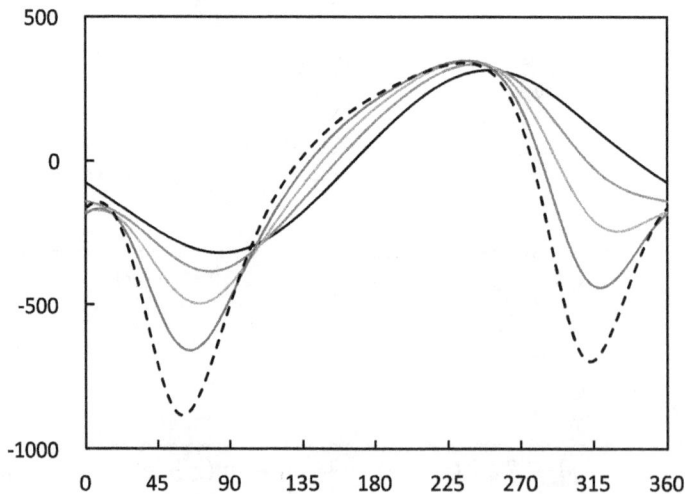

Figure 6.6. The scaled potential $\tilde{V}_1(\gamma_1)$ as a function of γ_1 for $\xi = 0.1$ (black), $\xi = 0.2$ (red), $\xi = 0.3$ (green), $\xi = 0.4$ (blues) and $\xi = 0.5$ (dashed).

If the water molecules are closer to each other, $\xi \overset{>}{\sim} 0.3$, the potential energy $\tilde{V}_1(\gamma_1)$ has *two* minima $\gamma_{1,\,min}$ and $\gamma^*_{1,\,min}$ corresponding to two stable configurations of the water dimer. This is shown in figure 6.7. In both configurations, one hydrogen atom is close to the line connecting the oxygen atoms, i.e. this is the *bonded* hydrogen atom. The minima are unequal, $\tilde{V}_{1,\,min} < \tilde{V}^*_{min}$, and is readily explained. We find that the more stable minima corresponds to the case when the unbounded hydrogen ion is farther away from the hydrogen ions of molecule #1. The minima and maxima of $\tilde{V}_1(\gamma_1)$ are listed in table 6.2. We note that the value $\xi \simeq 0.3$ corresponds to the empirical value of the separation between water molecules, $\xi_{expt} = 0.95/2.95 = 0.32$. This is a curious result and worthwhile to explore further in future work.

(2) We now relax the constraint $\gamma_2 = 0$ and allow molecule #2 to rotate about the y-axis. We calculate the variation of the electrostatic potential,

$$\Delta \tilde{V}_2(\gamma_2) = \tilde{V}(\gamma_{1,\,min}, \gamma_2) - \tilde{V}_{1,\,max}. \tag{6.26}$$

We thus have $\Delta \tilde{V}_2(0) = 0$. The function $\Delta \tilde{V}_2(\gamma_2)$ for five different values of the separation between molecules is shown in figure 6.8. For $0.1 \leqslant \xi < \xi \leqslant 0.4$, the potential difference has a minimum at $\gamma_2 = 0$ and a

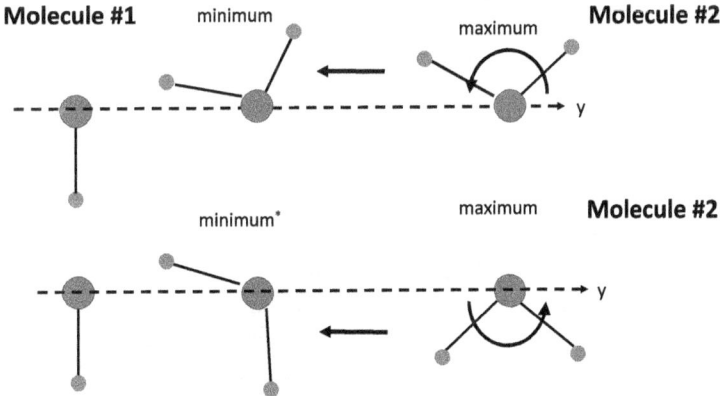

Figure 6.7. Water dimer for the minimum configurations corresponding to $\gamma_{1,\,min}$ and $\gamma^*_{1,\,min}$ for $\gamma_2 = 0$.

Table 6.2. The minima and maxima of the electrostatic potential energy $\tilde{V}_1(\gamma_1)$ for the five different values of the parameter $\xi = a/D$.

ξ	$\gamma_{1,\,min}$	$\tilde{V}_{1,\,min}$	$\gamma_{1,\,max}$	$\tilde{V}_{1,\,max}$	$\gamma^*_{1,\,min}$	$\tilde{V}^*_{1,\,min}$
0.1	84	−320.5	250	313.9	—	—
0.2	78	−384.8	242	334.6	—	—
0.3	70	−495.8	238	345.9	328	−244.7
0.4	66	−659.5	236	346.8	318	−439.7
0.5	60	−884.1	236	339.2	312	−696.8

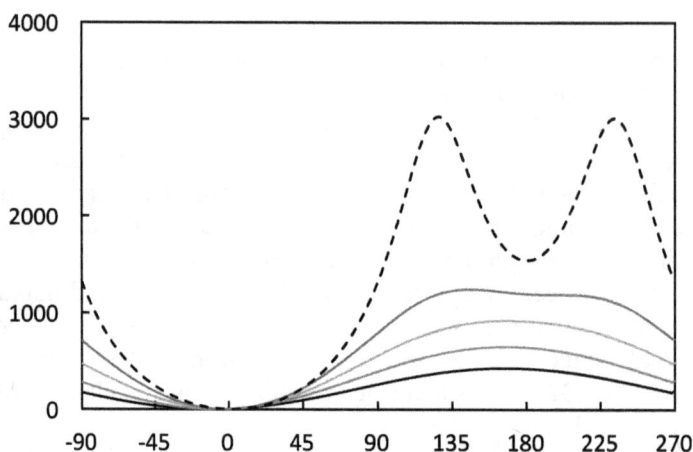

Figure 6.8. Potential energy difference $\Delta \tilde{V}_2(\gamma_2)$ as a function of γ_2 for $\gamma_1 = \gamma_{1, \min}$ for $\xi = 0.1$ (black), $\xi = 0.2$ (red), $\xi = 0.3$ (green), $\xi = 0.4$ (blue), and $\xi = 0.5$ (dashed).

maximum at $\gamma_2 = 180°$. For $\xi = 0.5$ the maximum near $\gamma_2 \simeq 135°$ and $\gamma_2 \simeq 225°$ corresponds to the case when the unbonded hydrogen of molecule #2 is opposite to either one of the two hydrogen ions of molecule #1.

We now choose the value $\xi = 0.3 \simeq \xi_{\text{expt}}$ and find the harmonic approximations of the potential energies; that is,

$$\tilde{V}_1(\gamma_1) \simeq \frac{1024}{2}(\gamma_1 - \gamma_{1, \min})^2, \tag{6.27}$$

$$\Delta \tilde{V}_2(\gamma_2) \simeq \frac{424}{2}\gamma_2^2. \tag{6.28}$$

Thus the torsional constants are $\kappa_1 = 1024$ and $\kappa_2 = 424$. The molecule #2 undergoes librational oscillations about the equilibrium configuration [1]. We note that a detailed comparison with spectroscopy data is not possible since our calculation assumes that molecule #1 is fixed and that the O–O distance is fixed, as well. The γ_1-libration involves the rotation of both the bonded and unbonded hydrogen atom. Since $m_H = 1$, the moment of inertia is $I_1 = 2R_{eq}^2$ or

$$I_1 = 2(0.95)^2 = 1.8. \tag{6.29}$$

The bonded hydrogen atom is along the axis of rotation for the γ_2-libration so that the moment of inertia follows, $I_2 = R_{eq}^2$, or

$$I_2 = 0.9. \tag{6.30}$$

We find the librational frequencies, $f_i = (2\pi)^{-1}\sqrt{\kappa_i/I_i}$, or

$$f_1 = 1260\text{cm}^{-1}, \qquad f_2 = 1790\text{cm}^{-1}, \tag{6.31}$$

and are thus part of the IR spectrum, cf. table 2.1.

6.5 Future directions

The goal of the book is to provide the reader with a non-technical introduction to the role of electrostatics in molecular systems. We focus on the properties of electrons in atoms, the nature of chemical bonds in molecules, and the formation of clusters via hydrogen bonds between molecules. A central goal was the explanation of the various energy scales and how it relates to various spectroscopic techniques. We describe the interactions between atoms and molecules in terms of electrostatic forces and repulsive forces that capture the quantum-mechanical repulsion of electrons. The balance between attractive and repulsive forces can be described by pseudoforces that can be modeled by harmonic springs.

The next step would be to take this description and apply it to problems in physical chemistry [1]. Intermolecular forces are central to problems from many areas of science [10] and, in particular, dominate the interactions of particles with surfaces [4]. In biological systems, molecular driving forces are dominated by electrostatic interactions [2, 8]. Simple models can be used to explain biological mechanisms in terms of pseudoforces, as we discussed in this text. The reader is invited to find simple models for these processes and to explore the role of electrostatic interactions.

References

[1] Berry R S, Rice S A and Ross J 2000 *Physical Chemistry* (New York: Oxford University Press)
[2] Dill K A and Bromberg S 2010 *Molecular Driving Forces* 2nd edn (London: Garland Science–Taylor and Francis)
[3] Harrison E R 1987 *Darkness at Night: A Riddle of the Universe* (Cambridge, MA: Harvard University Press)
[4] Israelachvili J 1992 *Intermolecular and Surface Forces* 2nd edn (London: Academic)
[5] Keutsch F N and Saykally R J 2001 Water clusters: Untangling the mysteries of liquids, one molecule at a time *Proc. Natl. Acad. Sci. U.S.A.* **98** 10533–40
[6] Landau L D and Lifshitz E M *Statistical Physics–Volume V of Course in Theoretical Physics* 3rd edn (Oxford: Butterworth and Heineman)
[7] Lindemann F 1910 The calculation of molecular vibration frequencies *Z. Phys.* **11** 1–18 (in German)
[8] Phillips R, Kondev J, Theriot J and Garcia H G 2003 *Physical Biology of the Cell* 2nd edn (London: Garland Science–Taylor and Francis)
[9] Sethna J P 2006 *Statistical Mechanics: Entropy, Order Parameters, and Complexity* (New York: Oxford University Press)
[10] Stone A J 2013 *Theory of Intermolecular Forces* 2nd edn (New York: Oxford University Press)
[11] Strogatz S 2003 *Sync – How Order Emerges from Chaos in the Universe, Nature, and Daily Life* (New York: Hachette Books)
[12] Zürcher U 2017 Electrostatics at the molecular level *Eur. J. Phys.* **38** 015206

www.ingramcontent.com/pod-product-compliance
Lightning Source LLC
Chambersburg PA
CBHW082105210326
41599CB00033B/6594